Effects of the Deletion of Chemical Agent Washout on Operations at the Blue Grass Chemical Agent Destruction Pilot Plant

Committee on Effects of the Deletion of Chemical Agent Washout
on Operations at the Blue Grass Chemical Agent Destruction Pilot Plant

Board on Army Science and Technology

Division on Engineering and Physical Sciences

The National Academies of
SCIENCES • ENGINEERING • MEDICINE

THE NATIONAL ACADEMIES PRESS
Washington, DC
www.nap.edu

THE NATIONAL ACADEMIES PRESS • 500 Fifth Street, NW • Washington, DC 20001

This activity was supported by Contract No. W911NF-15-1-0465 with the U.S. Department of Defense. Any opinions, findings, conclusions, or recommendations expressed in this publication do not necessarily reflect the views of any organization or agency that provided support for the project.

International Standard Book Number-13: 978-0-309-38948-8
International Standard Book Number-10: 0-309-38948-8
Digital Object Identifier: 10.17226/21884

Additional copies of this report are available for sale from the National Academies Press, 500 Fifth Street, NW, Keck 360, Washington, DC 20001; (800) 624-6242 or (202) 334-3313; http://www.nap.edu.

Copyright 2016 by the National Academy of Sciences. All rights reserved.

Printed in the United States of America

Suggested citation: National Academies of Sciences, Engineering, and Medicine. 2016. *Effects of the Deletion of Chemical Agent Washout on Operations at the Blue Grass Chemical Agent Destruction Pilot Plant*. Washington, DC: The National Academies Press. doi:10.17226/21884.

The National Academies of
SCIENCES · ENGINEERING · MEDICINE

The **National Academy of Sciences** was established in 1863 by an Act of Congress, signed by President Lincoln, as a private, nongovernmental institution to advise the nation on issues related to science and technology. Members are elected by their peers for outstanding contributions to research. Dr. Ralph J. Cicerone is president.

The **National Academy of Engineering** was established in 1964 under the charter of the National Academy of Sciences to bring the practices of engineering to advising the nation. Members are elected by their peers for extraordinary contributions to engineering. Dr. C. D. Mote, Jr., is president.

The **National Academy of Medicine** (formerly the Institute of Medicine) was established in 1970 under the charter of the National Academy of Sciences to advise the nation on medical and health issues. Members are elected by their peers for distinguished contributions to medicine and health. Dr. Victor J. Dzau is president.

The three Academies work together as the **National Academies of Sciences, Engineering, and Medicine** to provide independent, objective analysis and advice to the nation and conduct other activities to solve complex problems and inform public policy decisions. The Academies also encourage education and research, recognize outstanding contributions to knowledge, and increase public understanding in matters of science, engineering, and medicine.

Learn more about the National Academies of Sciences, Engineering, and Medicine at **www.national-academies.org.**

COMMITTEE ON EFFECTS OF THE DELETION OF CHEMICAL AGENT WASHOUT ON OPERATIONS AT THE BLUE GRASS CHEMICAL AGENT DESTRUCTION PILOT PLANT

GARY S. GROENEWOLD, Idaho National Laboratory, Idaho Falls, *Chair*
HEREK L. CLACK, University of Michigan, Ann Arbor
RICHARD C. FLAGAN, NAE,[1] California Institute of Technology, Pasadena
REBECCA A. HAFFENDEN, Argonne National Laboratory, Santa Fe, New Mexico
THOM J. HODGSON, NAE, North Carolina State University, Raleigh, North Carolina
MURRAY GLENN LORD, The Dow Chemical Company, Freeport, Texas
WILLIAM J. WARD, NAE, GE Corporate Research and Development (retired),
    Niskayuna, New York

Staff

BRUCE BRAUN, Director, Board on Army Science and Technology
JAMES C. MYSKA, Study Director
NIA D. JOHNSON, Senior Research Associate
DEANNA SPARGER, Program Administrative Coordinator

---

[1] Member of the National Academy of Engineering.

BOARD ON ARMY SCIENCE AND TECHNOLOGY

DAVID M. MADDOX, NAE,[1] Independent Consultant, Arlington, Virginia, *Chair*
JEAN D. REED, National Defense University, Arlington, Virginia, *Vice Chair*
SCOTT BADENOCH, Badenoch, LLC, Southfield, Michigan
STEVEN W. BOUTELLE, CISCO Consulting Services, Herndon, Virginia
CARL A. CASTRO, Center for Innovation and Research and Military Families, University of Southern California, Los Angeles
DAVID E. CROW, NAE, University of Connecticut, Glastonbury
REGINALD DESROCHES, Georgia Institute of Technology, Atlanta
EARL H. DOWELL, NAE, Duke University, Durham, North Carolina
FRANCIS J. DOYLE III, Harvard University, Cambridge, Massachusetts
JULIA D. ERDLEY, Pennsylvania State University, State College
LESTER A. FOSTER, Electronic Warfare Associates, Herndon, Virginia
JAMES A. FREEBERSYSER, BBN Technology, St. Louis Park, Minnesota
PETER N. FULLER, Cypress International, Springfield, Virginia
R. JOHN HANSMAN, NAE, Massachusetts Institute of Technology, Cambridge
J. SEAN HUMBERT, University of Colorado, Boulder
JOHN W. HUTCHINSON, NAE/NAS,[2] Harvard University, Cambridge, Massachusetts
JENNIE HWANG, NAE, H-Technologies Group, Cleveland, Ohio
BRUCE D. JETTE, Synovision Solutions, LLC, Burke, Virginia
JOHN JOANNOPOULOS, NAS, Massachusetts Institute of Technology, Cambridge
ROBIN L. KEESEE, Joint Improvised Explosive Device Defeat Organization (retired), Fairfax, Virginia
ERIC T. MATSON, Purdue University, West Lafayette, Indiana
ROGER L. McCARTHY, NAE, McCarthy Engineering, Palo Alto, California
MICHAEL McGRATH, McGrath Analytics, LLC, Reston, Virginia
ALLAN T. MENSE, Raytheon Missile Systems, Tucson, Arizona
WALTER F. MORRISON, Booz Allen Hamilton (retired), Alexandria, Virginia
SCOTT PARAZYNSKI, Arizon State University, Tempe
DANIEL PODOLSKY, NAE, University of Texas Southwestern Medical Center, Dallas
KENNETH M. ROSEN, General Aero-Science Consultants, LLC, Guilford, Connecticut
LEON E. SALOMON, Independent Consultant, Gulfport, Florida
ALBERT A. SCIARRETTA, CNS Technologies, Inc., Springfield, Virginia
NEIL SIEGEL, NAE, North Grumman Information Systems, Carson, California
LAWRENCE D. STONE, NAE, Metron, Inc., Reston, Virginia
MICHAEL A. VANE, Independent Consultant, Shaver Lake, California

Staff

BRUCE A. BRAUN, Director
CHRIS JONES, Financial Associate
JAMES C. MYSKA, Program Officer
NIA D. JOHNSON, Senior Research Associate
DEANNA P. SPARGER, Program Administrative Coordinator

---

[1] Member of the National Academy of Engineering.
[2] Member of the National Academy of Sciences.

Preface

The Blue Grass Chemical Agent Destruction Pilot Plant (BGCAPP) was designed and constructed at the Blue Grass Army Depot in Richmond, Kentucky, for the purpose of destroying rockets and artillery projectiles that contain the nerve agents GB and VX. These nerve agents are chemical warfare agents, and the United States is obligated by the Chemical Weapons Convention treaty to destroy them. BGCAPP is a chemical processing plant specifically designed to access munition cavities and drain liquid agents. Agent remaining in the munitions was originally to have been washed out using a stream of hot, high-pressure water. The agent and washout water were to have been combined and then reacted with sodium hydroxide, which chemically degrades the nerve agents. In addition to treating the agent, the BGCAPP also processes other multiple solid, liquid, and gaseous waste streams.

The mixtures of the agents with washout water generated by the initial drain-water washout process had the potential to create problems that would degrade the safety of operations and compromise the materials used in the agent transfer lines. Consequently, BGCAPP program management decided that these problems could be mitigated by eliminating the water washout process. This decision resulted in munitions bodies containing significant residual agent on surfaces or in crevices being sent into some BGCAPP processes that, in some instances, were not originally designed for agent destruction; and in others, resulted in higher agent loads than originally intended.

This report describes possible outcomes of the deletion of the water washout process that are related to the partitioning of agent into multiple processing streams within BGCAPP. These outcomes include the necessity of some processing units treating more agent than was initially planned, impacts on BGCAPP's ability to meet and demonstrate achieving legally required destruction efficiency criteria, and impacts on process modeling and the ability to computationally predict munitions throughput and completion dates for munitions destruction campaigns.

I am very thankful for the members of the Committee on Effects of the Deletion of Chemical Agent Washout on Operations at the Blue Grass Chemical Agent Destruction Pilot Plant, who served in a volunteer capacity but nevertheless were exceptionally generous with their expertise and time. They attended briefings at BGCAPP and two writing meetings at the National Academies of Sciences, Engineering, and Medicine facilities in Washington, D.C.

The committee is indebted to the BGCAPP staff, being the beneficiary of extensive briefings and literature that they provided. The BGCAPP staff members were remarkable in their patience and energy as they responded to repeated requests for information. They displayed a high level of expertise throughout the course of this study.

The committee is also grateful for the support of the Academies staff, particularly Deanna Sparger, Nia Johnson, Jim Myska, and Bruce Braun. Their attention to logistical detail and the long-running familiarity with the BGCAPP endeavor was significant and highly appreciated over the course of this study.

Gary S. Groenewold, *Chair*
Committee on Effects of the
Deletion of Chemical Agent Washout on
Operations at the Blue Grass Chemical
Agent Destruction Pilot Plant

Acknowledgments

This report has been reviewed in draft form by individuals chosen for their diverse perspectives and technical expertise, in accordance with procedures approved by the Report Review Committee. The purpose of this independent review is to provide candid and critical comments that will assist the institution in making its published report as sound as possible and to ensure that the report meets institutional standards for objectivity, evidence, and responsiveness to the study charge. The review comments and draft manuscript remain confidential to protect the integrity of the deliberative process. We wish to thank the following individuals for their review of this report:

Edward L. Cussler, Jr., University of Minnesota, Minneapolis,
Jaron Hansen, Brigham Young University,
Todd Kimmell, Argonne National Laboratory,
Ronald Kolpa, Argonne National Laboratory,
Arturo Lopez, The Dow Chemical Company,
Trisha H. Miller, Sandia National Laboratories, and
Julius Rebek, Jr., The Scripps Research Institute.

Although the reviewers listed above have provided many constructive comments and suggestions, they were not asked to endorse the conclusions or recommendations nor did they see the final draft of the report before its release. The review of this report was overseen by Hyla S. Napadensky, Napadensky Energetics Inc. (retired), who was responsible for making certain that an independent examination of this report was carried out in accordance with institutional procedures and that all review comments were carefully considered. Responsibility for the final content of this report rests entirely with the authoring committee and the institution.

Contents

SUMMARY 1

1 INTRODUCTION 7
    Processing of Rocket Warheads, 7
    Processing of Projectiles, 10
    Concerns Related to the Operation of the Plant as Originally Designed, 10
    Washout Deletion Modifications, 12
    Statement of Task, 12
    Approach to the Statement of Task and Organization of the Report, 12
    References, 12

2 PLANT PROCESS CHANGES AS A RESULT OF WASHOUT DELETION 13
    Rocket and Projectile Drain Operations, 13
    Strainer Sock Loading, 14
    MPT and Plant Off-Gas Treatment System (OTM), 15
    EBH and Off-Gas Treatment System for the Energetics Neutralization System (OTE), 15
    Impacts on the Overall MDB HVAC System, 16
    References, 16

3 IMPACTS ON CALCULATION OF DESTRUCTION EFFICIENCY 18
    Current Regulatory Requirements for Destruction Efficiency, 18
    Current Approach to Calculating Destruction Efficiency (Approach 1), 19
    Impact of Washout Deletion on the Calculation of Destruction Efficiency, 19
        Agent Partitioning to the EBH and ENS Units, 21
        Agent Partitioning to the MPT and OTM Units, 23
    Alternative Approaches to Calculating Destruction Efficiency (Approaches 2 and 3), 24
        Calculation of Destruction Efficiency Under Approach 2, 25
        Calculation of Destruction Efficiency Under Approach 3, 25
    Measurements Required for Verifying Destruction Efficiency Requirements, 28

4 PROCESS MODELING IN SUPPORT OF WASHOUT DELETION 29
    The BGCAPP Facility Model, 29
        Munition Drain Times, 30
        Filter Sock Change-Out Frequency, 30
    Exploring System Sensitivity to the Input Parameters, 30
    Operational Data Collection, 31
    Reference, 32

APPENDIXES

A Committee Activities   35
B Biographical Sketches of Committee Members   36

Figures and Table

FIGURES

1-1 Block diagram showing the processing units and the flow of products at BGCAPP, 8

2-1 Agent transfer system block diagram, 14
2-2 Worst-case GB scenario with venting to room, 17

3-1 Process flow diagram for destruction efficiency calculation under Approaches 1, 2, and 3, 20
3-2 Flow diagram showing committee recommendations for expanding effluent measurements to allow the calculation of DE at 99.9999 regulatory requirements and for rerouting the OTE through the OTM, 26

TABLE

3-1 Summary of Approaches to Calculating Destruction Efficiency (DE), 27

Abbreviations and Acronyms

ACS	agent collection system
AFS	aluminum filtration system
ANR	agent neutralization reactor
ANS	agent neutralization system
APS	aluminum precipitation system
ATT	agent transfer tank
BGAD	Blue Grass Army Depot
BGCAPP	Blue Grass Chemical Agent Destruction Pilot Plant
BPBG	Bechtel Parsons Blue Grass
BPT	Bechtel Pueblo Team
CWC	Chemical Weapons Convention
DE	destruction efficiency
EBH	energetics batch hydrolyzer
ENR	energetics neutralization reactor
ENS	energetics neutralization system
FOAK	first-of-a-kind
GB	nerve agent (sarin)
HF	hydrofluoric acid or hydrogen fluoride
HSA	hydrolysate storage area
HVAC	heating, ventilation, and air conditioning
KAR	Kentucky Administrative Regulations
KDEP	Kentucky Department for Environmental Protection
KRS	Kentucky Revised Statutes
LT	level transmitter
MDB	munitions demilitarization building
MPT	metal parts treater
MWS	munitions washout system
OTE	EBH off-gas treatment system
OTM	off-gas treatment system
PEO ACWA	Program Executive Office, Assembled Chemical Weapons Alternatives
PHS	projectile handling system
ppb	parts per billion
RCM	rocket cutting machine
RCRA	Resource Conservation and Recovery Act
RD&D	Research, Development, and Demonstration
RHA	residue handling area
RHS	rocket handling system
RO	reverse osmosis
RSM	rocket shear machine
SCWO	supercritical water oxidation
SDS	spent decontamination solution
TOX	thermal oxidizer
VX	nerve agent

Summary

The United States has signed and ratified the Chemical Weapons Convention, which outlaws the production and possession of chemical weapons and a number of related chemicals. To date, the United States has destroyed about 90 percent of its stockpile, mostly using incineration.

As part of the U.S. effort to destroy its remaining stockpile of chemical munitions, the Department of Defense is building the Blue Grass Chemical Agent Destruction Pilot Plant (BGCAPP) on the Blue Grass Army Depot (BGAD), near Richmond, Kentucky. The stockpile stored at BGAD consists of rockets and projectiles containing the nerve agents GB and VX and the blister agent mustard. Continued storage poses a risk to the BGAD workforce and the surrounding community because these munitions are several decades old and are developing leaks. The projectiles containing mustard agent will be destroyed using a Static Detonation Chamber being built adjacent to BGCAPP. BGCAPP will destroy the rockets and projectiles containing GB and VX. The variety of munition and agent types, and the degrading agent they contain, poses a variety of challenges to their destruction.

Due to public opposition to the use of incineration to destroy the BGAD stockpile, Congress mandated that non-incineration technologies be identified for use at BGCAPP.[1] As a result, BGCAPP will destroy the GB and VX by hydrolysis using hot caustic solution (sodium hydroxide). To comply with the Chemical Weapons Convention requirements for the destruction of chemical weapons,[2] the resulting hydrolysates must be further treated. At BGCAPP, this will be accomplished using supercritical water oxidation.

The original BGCAPP design called for munitions to be drained of agent and then for the munition bodies to be washed out using high-pressure hot water. However, during the course of committee discussions related to the systemization of BGCAPP, several concerns emerged that held the potential to compromise safe operations and impede agent processing throughput in the plant. Much of the concern focused on the mixture of agent and wash water that was produced during agent drain and water washout operations. The mixing water and VX has the potential to cause an autocatalytic exothermic reaction that can lead to frothing and overflow in storage tanks upstream of the agent neutralization reactors. The storage tanks are not designed to contain the reaction, in contrast to the neutralization reactors, which are designed to operate at high temperature. Water and VX mixtures can also produce agent gels that could impact agent destruction processes. Mixing water and GB can produce significant hydrofluoric acid, which can be damaging to the steel transfer lines between tanks.

As a result, as part of a larger package of modifications called Engineering Change Proposal 87 (ECP-87), the munition washout step was eliminated. However, implementing this solution will cause larger quantities of agent—more than originally planned—to be partitioned into different BGCAPP processes where agent destruction for these larger quantities is unproven because those processes were designed to treat only small amounts of residual agent. This could have the unintended effect of compromising the ability of the plant to achieve and demonstrate the Kentucky statutory requirement of 99.9999 percent destruction efficiency (DE).

The Program Executive Officer for Assembled Chemical Weapons Alternatives asked that an ad hoc study committee be formed to look into the effects of deleting the water washout step. The statement of task of the Committee on Effects of the Deletion of Chemical Agent Washout on Operations at the Blue Grass Chemical Agent Destruction Pilot Plant was as follows:

---

[1] A similar neutralization plant is also being completed at Pueblo Chemical Depot in Pueblo, Colorado.

[2] *Destruction of chemical weapons* means a process by which chemicals are converted in an essentially irreversible way to a form unsuitable for production of chemical weapons and which, in an irreversible manner, renders munitions and other devises unusable as such (Chemical Weapons Convention, Annex on Implementation and Verification, Part IV (A), Destruction of Chemical Weapons and Its Verification Pursuant to Article IV).

- Assess the impact of the design change on plant operations and the impacts to plant throughput, taking into account revised rocket and projectile drain times, strainer change-out frequency, and metal parts treater throughput;
- Review and assess the calculations associated with the ability of the metal parts treater and thermal oxidizer to effectively process additional residual agent GB and VX contained in the drained rocket and projectile munition bodies;
- Review and assess the contractor's approach to the destruction efficiency (DE) calculations and provide any suggestions that support the DE confirmation process; and,
- Assess the validity of process modeling conducted to date and recommend where additional modeling may be of benefit for understanding likely plant operation performance.

BGCAPP is legally required to achieve what is termed "six-nines" destruction of GB and VX, which means that it must be demonstrated that the fraction of agent destroyed be greater than 0.999999, or alternatively, the fraction remaining must be less than 1×10^{-6} of what was originally present in the munitions. In the original plant design, it was intended that almost all of the agent would be processed by caustic hydrolysis through the agent neutralization system (ANS). Of course, this was never strictly the case because some vapors from the agent draining operations will be vented into the munitions demilitarization building (MDB) heating, ventilation, and air conditioning (HVAC) filtration system, and there was always likely to be traces of agent remaining on the projectiles and rocket warheads after the agent washout. But these quantities were deemed to be negligible, and thus DE could be demonstrated by merely measuring agent in the ANS hydrolysate to a concentration equivalent to or less than 1×10^{-6} times that of the concentration of agent fed into the unit.

The deletion of the water washout step will now result in significant quantities of agent being partitioned into other process streams of BGCAPP. The rocket warhead pieces (after the warheads are drained and sheared) will contain more residual agent than originally planned when they are sent to the energetics batch hydrolyzer (EBH) units. These units, designed to hydrolyze the energetic materials in the rocket warhead bursters, also contain caustic (sodium hydroxide) at elevated temperature. Calculations by BGCAPP predict that there will be sufficient excess caustic present to ensure complete destruction of any agent that is partitioned to the EBHs, and this expectation is supported by prior experience with caustic hydrolysis of GB and VX. GB, on account of its solubility, will very likely be completely eliminated from the caustic solution. However, VX may survive as a result of incomplete mixing or as a result of being sequestered in cracks in the metal parts. Destruction efficiency of agent in the EBHs is not explicitly known and will need to be demonstrated in order to provide a defensible calculation of DE. In addition, there is also a chance that a fraction of the agent, particularly GB, which has a lower boiling point, will be partitioned into the off-gas stream from the EBHs into the EBH off-gas treatment system (OTE). Significantly, the OTE is not designed to destroy agent, which means that any agent that is volatilized in the EBHs will instead be captured on the carbon filters of the MDB HVAC. While extensive experience in the broader chemical demilitarization program indicates that this outcome would be protective of the public and the environment, BGCAPP believes that it is not likely to be allowed to take credit for the removal of agent vapor by the carbon bank adsorption prior to release of exhaust into the atmosphere when DE is calculated per the Kentucky Revised Statutes or the BGCAPP operating permit.[3]

The drained projectile bodies will also contain more residual agent than originally planned. Because these items do not have energetics components, they will be processed through the metal parts treater (MPT). The MPT thermally decontaminates agent-contaminated items by ensuring that they are exposed to 1,000°F for at least 15 minutes. Extensive operational experience and calculations by BGCAPP indicate that processing time and temperature of the MPT should be sufficient to destroy any extra agent partitioned into that unit, but, as in the case of the EBHs, this will not be known until the unit is actually operated. The gaseous effluent from the MPT flows into the off-gas treatment system (OTM), which is equipped with a thermal oxidizer (TOX), and the committee believes that the TOX will destroy any fugitive agent vapors escaping from the MPT.

Demonstrating six-nines destruction after deletion of the washout step will be significantly more difficult than originally planned due to the change in the amount of agent now partitioned outside of the ANS. BGCAPP personnel have considered two alternative methodologies to determine DE, but these entail much more measurement, and, in many cases, good analytical methods do not currently exist. With some streams, like the caustic in the EBHs, it may be difficult to measure agent concentration down to a level that would demonstrate achievement of the DE criteria.

An attractive solution would be to count the agent trapped in the MDB HVAC carbon beds as destroyed. However, the Kentucky Department for Environmental Protection has stated that BGCAPP may not take credit for measurements downstream of the carbon filtration system without revision to the statute/guidelines; thus, agent trapped in the MDB HVAC carbon beds cannot be counted as destruction in the calculation of DE, even though the design and build of this system is consistent with the capture/removal devices approved for use in other (incineration) demilitarization

---

[3] John McArthur, environmental manager, BPBG, "Destruction Efficiency Considerations," presentation to the committee on September 9, 2015.

facilities where removal was considered in the calculation of "destruction and/or removal efficiency" at those sites.[4]

These overarching assessments are summarized in a series of findings and recommendations, which provide a summary of all of the committee's work. Chapter 2 assesses the process impacts of washout deletion. Specific findings and recommendations are focused on the effects of agent degradation over decades in storage on the physical state of the agent and, hence, agent drain times. A second issue is that the change-out of the filter socks used to capture agent solids may be time consuming, even though new socks with greater capacity have been introduced, and, with washout deletion, the change-outs will demand additional time, which has not been included in the process modeling. Additional attention was focused on the potential for the impact of increased agent loading to the OTM, although there are approaches for ensuring a high DE. Similarly, the EBHs and units situated serially downstream will be subjected to additional agent loading. The effect of the washout deletion on agent loading to the MDB HVAC system is uncertain at this time. It is likely that there will be additional agent loading from the EBH-OTE process stream. In addition, there may be a change in the amount of agent directed to the MDB HVAC from the munitions-draining operations.

Finding 2-1. Uncertainty in the number of munitions containing degraded agent and the degree of agent degradation is compounded by a lack of knowledge of the physico-chemical characteristics of degraded agent as they relate to drain times and amounts of residual agent retained in munitions at the end of the drain process. Better data are needed to properly estimate the time that will be required to process the nerve agent munitions through BGCAPP.

Recommendation 2-1. BGCAPP should gather data, such as mass drained, drain time, and any available information on physical state, for each individual munition during operations ramp up to assess the state of the agent fills and thus expected variability in drain times for each agent lot and type of munition. The acquisition of these data should continue throughout operations to continuously improve the quality of estimates as an aid toward planning of plant operations and to estimate completion times.

Finding 2-2. Even with the change in filter sock capacity, the change-out frequency could become the rate-determining step in the processing of rockets and projectiles.

Finding 2-3. Agent processed through the MPT and the off-gas treatment system will constitute a significant fraction of the agent destroyed at BGCAPP. This is a departure from the original design where almost the entire agent volume was being treated by hydrolysis.

Finding 2-4. Multiple mechanisms exist for controlling the MPT throughput rate to reduce instantaneous agent loading in the MPT and the off-gas treatment system. These include, but are not limited to, approaches such as increasing the residence time in zone 1 of the MPT, reducing the number of projectiles on each tray being processed and increasing the steam addition rate to the MPT.

Recommendation 2-2. BGCAPP should evaluate whether higher agent vaporization rates in the metal parts treater (MPT) can be accommodated by optimizing the operating parameters of the MPT, the off-gas treatment system, and associated systems.

Finding 2-5. With the deletion of munitions washout, some of the chemical agent from the rocket warheads will be sent to the EBHs. Some fraction of the agent introduced into the EBHs will be volatilized and then flow into the EBH OTE. The OTE does not have a TOX, so some of the agent transported from the EBH to the OTE may penetrate to the MDB HVAC.

Recommendation 2-3. BGCAPP should conduct modeling and experimental studies to bound the quantity of agent present in the OTE vent stream (stream #8517).

Finding 2-6. During punch and drain operations, vapors are released directly to the room air and are exhausted through the MDB HVAC system. The primary mode of capture of these vapors is the carbon filter bank. This function is part of the original plant process; however, the washout deletion may affect agent concentrations in the gas phase that will be transferred to the MDB HVAC system.

Recommendation 2-4. BGCAPP should complete modeling to estimate the agent load to the carbon beds in the absence of a munition washout step to ensure that the lifetime of these beds is known.

As noted above, the washout deletion will have a pronounced effect on the calculation of DE that is mandated by Kentucky Revised Statutes; this is the subject of Chapter 3. Although some findings and recommendations in Chapter 3 overlap with those offered in Chapter 2, the focus in Chapter 3 is on the potential for washout deletion to complicate the calculation of DE and supporting measurements. Specifically, because an increased fraction of agent will now be partitioned into the EBHs, a possible pathway for agent would be any agent residual on the metal parts following processing by the EBHs, which would then be transferred into the MPT. Given the harsh treatment conditions in the EBH and MPT, it is not likely that agent would survive these

---

[4] NRC Washout Deletion Committee Questions and Responses 151029, received via e-mail on December 8, 2015.

units; however, this has not been demonstrated. Additionally, the EBHs generate energetics hydrolysate. This hydrolysate is then sent to the energetics neutralization system (ENS), which operates at a higher temperature and pressure than the EBHs. BGCAPP calculations indicate that any residual agent in the hydrolysate should be destroyed by the ENS, but, again, this has not been demonstrated for the additional loading that could result from washout deletion. The off-gas from the EBHs may also contain some agent. This off-gas will be treated by the OTE, which is not designed to destroy agent, and its capacity to do so is unknown.

After washout deletion, the facility OTM will need to accommodate more agent than was originally intended. It is probably capable of doing this, but BGCAPP will need to demonstrate this in order to provide assurance of DE. The particles formed in the TOX unit of the OTM will very likely be free of agent, as a consequence of the high temperature and residence time of the agent in the TOX. However, BGCAPP will need to demonstrate DE for this solid waste stream from the TOX.

To account for the possibility of agent in these non-ANS effluent streams, BGCAPP has considered two possible alternative approaches to the original approach (called Approach 1) for demonstrating DRE. The first alternative, Approach 2, would measure the difference in agent quantities in the feed and effluent streams from specific individual units. The second alternative, Approach 3, would assume a bulk quantity in the munitions input lines, while measuring agent quantities in the effluent streams from all treatment units. As stated in the findings below, Approach 2 is not viable for demonstrating DE criteria because it does not include all possible agent-contaminated streams. Approach 3 is more achievable but would present significant challenges in developing measurement methodologies for the different forms of effluent.

One recommendation related to the physical plant did emerge from the committee's extended DE discussion. It was noted that because the OTE was not designed to destroy organics (such as agent), a possible modification might be to send the off-gas from the OTE into the OTM, which is designed to destroy organics and is expected to be able to destroy agent.

Finding 3-1. With washout deletion, the current methodology for calculating whether BGCAPP meets the statutory and regulatory requirement for a DE of 99.9999 percent will no longer be applicable. This is because the drained and washed out agent will no longer go only to the ANS. Instead, a greater amount of residual agent remaining in the rocket warhead pieces will now be processed in the EBHs and the ENS, and residual agent in the projectiles and possibly on rocket warhead pieces will now be processed through the MPT.

Finding 3-2. The partitioning of agent across additional process streams will introduce monitoring challenges that BGCAPP needs to be aware of. It may be difficult to identify monitoring technologies or strategies of sufficient sensitivity to measure what are expected to be low concentrations of agent in some streams. Additionally, the identification of new technologies or strategies carries the risk of a negative impact on the BGCAPP schedule.

Finding 3-3. It is likely that the combination of the EBH and MPT treatment conditions will be sufficient to destroy all residual agent residing on rocket warhead pieces. This, however, needs to be shown through either calculations or demonstration.

Recommendation 3-1. BGCAPP should calculate or otherwise demonstrate a 99.9999 percent ("six-nines") destruction efficiency (DE) for residual agent residing on rocket warhead pieces exiting the metals parts treatment unit (MPT). This would provide assurance that the solid effluent from the MPT (stream #7652) generated during rocket campaigns is free of agent to ensure compliance with DE requirements.

Finding 3-4. Complete destruction of augmented agent loadings passing through the EBH/ENS system has not been demonstrated.

Recommendation 3-2. BGCAPP should demonstrate satisfactory destruction efficiency for agent serially treated with caustic under the same conditions as those present in the energetics batch hydrolyzers (EBHs) and the energetics neutralization system at agent loadings equivalent to the highest quantities anticipated to be treated by the EBHs without washout.

Finding 3-5. An unknown fraction of agent entering the EBHs during the rocket campaigns may undergo volatilization instead of hydrolysis. Volatilized agent will be processed through the OTE system (stream #8517), which is not designed to destroy agent. Agent escaping the OTE will be removed to the MDB HVAC carbon filter banks, together with fugitive agent emissions from the munition drain processes. Because agent partitioned into these pathways cannot be counted as destroyed, and because BGCAPP believes it is not likely to be allowed to take credit for removal of agent vapor by carbon bank adsorption in the MDB HVAC prior to release of exhaust to the atmosphere, the implementation of washout deletion will require significant permit modifications and has the potential to prevent BGCAPP from achieving DE criteria.

Recommendation 3-3. For all of the gaseous process streams, BGCAPP should rigorously demonstrate that negligible agent is partitioned into the munitions demilitarization building (MDB) heating, ventilation, and air conditioning (HVAC) carbon filter banks under all conditions that could arise during the rocket campaign. BGCAPP should provide for monitoring of the OTE effluent stream (#8517) with analytical sensitivity sufficient to ensure that destruction efficiency criteria are achieved before they enter the MDB HVAC system.

Recommendation 3-4. BGCAPP should examine the possibility of routing the gaseous effluent from the OTE (energetics batch hydrolyzer off-gas treatment system) into the OTM (off-gas treatment system). This would eliminate the biggest uncertainties in M_{Out} exiting the munitions demilitarization building (MDB), because it is likely that any agent surviving the OTE would be destroyed in the OTM. The number of gaseous streams from processing units exiting the MDB would be reduced to a single stream—namely, the off-gas from the OTM—and would be less likely to contain significant agent as a result of off-gas passing through the thermal oxidizer.

Finding 3-6. It is likely that the combination of the MPT and the OTM will completely destroy any agent entering the MPT. However, after washout deletion, the OTM will receive gaseous streams from other sources that may contain more agent than originally planned. It is currently unknown whether the OTM can adequately treat the combined load of all streams after washout deletion.

Recommendation 3-5. BGCAPP should measure solid, gaseous, and liquid effluents from the OTM (off-gas treatment system) during initial projectile campaigns to ensure that these effluents meet the destruction efficiency criteria.

Finding 3-7. The solid waste stream from the OTM should be agent-free. This conclusion will need to be demonstrated to the Kentucky Department for Environmental Protection based on validated process controls and statistical testing.

Finding 3-8. Approach 2 is not an appropriate option for the calculation of DE. It is incomplete because it does not include the gaseous emissions from the OTE—which, under the new configuration, may contain agent—and because it is not operationally practical to measure agent quantities in the feed to, and effluent from, the individual process units.

Finding 3-9. Approach 3 could conceivably be used for a defendable DE determination, because it accounts for the OTE gaseous process stream #8517, provided it is modified to include the fugitive releases of agent vapor directed to the MDB HVAC system. However, Approach 3 would require development of additional methodologies for measuring masses of agent partitioned into the two gaseous waste streams entering the MDB HVAC system.

Recommendation 3-6. If Approach 3 is adopted, then BGCAPP should evaluate the concentrations of agent liable to be present in all gaseous process streams and develop measurement approaches with sufficient sensitivity to ensure that destruction efficiency criteria are being achieved.

Finding 3-10. The performance requirements for the analytical measurement methodology for measuring agent in the off-gas process stream from the OTE (#8517) are not known, because the fraction of agent that will be partitioned into this stream is uncertain.

Recommendation 3-7. If Recommendation 3-4 is not pursued, BGCAPP should conduct research to determine what fraction of GB agent might partition into the off-gas process stream from the OTE (energetics batch hydrolyzer off-gas treatment system) and then use this information to set analytical performance requirements that can be used to identify analytical measurement methodology.

Chapter 4 addresses the modeling of munitions throughput at BGCAPP, which is done using an overall process model (rather than individual process models). Input parameters were based on point observations made by BGCAPP staff, which may well be accurate in the mean. However, the committee does not believe that the expected variability in plant operational parameters is reflected in the model.

An area of repeated concern for BGCAPP and for the committee was the accuracy of estimates of the time required to drain the munitions. Inaccurate drain-time estimates have the potential to result in inaccurate model throughput estimates. The committee believes that the best approach for estimating drain times would be to capture information on munition drain operations from individuals who have actually conducted these activities at other demilitarization sites. There may also be opportunities to collect actual operating data during systemization and as plant operations begin at BGCAPP. Another process concern is that of filter sock change-out, which is related to the issue of draining in that both processes are affected by the extent of solids and gels in the munitions. In general, BGCAPP staff have attempted to be conservative in all of their parameter estimates, but it is also clear that many of the parameters potentially have large variances (e.g., the amount of agent fill that cannot be drained). With regard to bounding the output of the model, it would be useful to run the model with input parameters equivalent to the highest and lowest levels that could be encountered. Finally, statistical quality control might have significant utility for managing operations.

Finding 4-1. While the process model explores the influence of variations in operating parameters on the performance of BGCAPP, the limited treatment of the stochastic nature of those parameters does not reflect operational experience.

Finding 4-2. The reliance on point estimates in the model data does raise concerns about the ability of the model to accurately forecast future facility operations in terms of the length of time to complete the processing of the chemical weapons and the risks involved in operating the facility.

Finding 4-3. The stochastic nature of the gelling or crystallization of the GB agent may still be partially retrievable. A formal debriefing of individuals who have drained munitions to capture the (informal and clearly anecdotal) nature of the condition of the agent in the weapons might be useful in developing more believable assumptions as to the condition and variability of the chemical agents in the weapons.

Recommendation 4-1. BGCAPP should retrieve and document historical (informal and anecdotal) data on munition drain times and run these data, complete with ranges of uncertainty, through the BGCAPP model.

Finding 4-4. The actual filter sock change-out rate may be the most important rate-limiting factor in BGCAPP operations and may be underestimated.

Finding 4-5. Analysis of the sensitivity of the BGCAPP operations to variations in model input parameters might expose potential operational issues, allowing them to be quantified and possibly mitigated prior to operations.

Recommendation 4-2. BGCAPP should design and execute a series of modeling experiments to determine the sensitivity of operations to variations in operating parameters, reflecting the stochastic nature of some processes. Examples of parameters include maintenance and repair times, added characterization steps, retreatment for batches not meeting destruction efficiency, and compounding problems such as long munitions drain times together with very frequent filter sock change-outs. The results of these experiments should be used to prepare for potential challenges and mitigate them ahead of time as much as possible.

Finding 4-6. Point estimates of operational parameters are only a starting point. To fully understand the plant operation and, ultimately, to understand the plant timeline, one needs data on the distribution of parameter values that may be encountered during operation.

Recommendation 4-3. During start-up, and continuing through plant operations, BGCAPP should gather data for relevant model parameters with sufficient resolution to assess the probability density functions for these parameters.

Finding 4-7. Statistical quality control could be a useful management tool for understanding and identifying possible problems as they occur.

Recommendation 4-4. BGCAPP should give attention to developing analysis tools such as statistical quality control prior to actual facility start-up.

1

Introduction

The United States manufactured significant quantities of chemical weapons during the Cold War and years prior. Because the chemical weapons are aging, storage constitutes an ongoing risk to the facility workforces and to the communities nearby. In addition, the Chemical Weapons Convention (CWC) treaty, which the United States has ratified, stipulates that the chemical weapons be destroyed. The United States has destroyed approximately 90 percent of the chemical weapons stockpile located at seven sites.[1] However, there are remaining stockpiles at the Blue Grass Army Depot (BGAD) in Richmond, Kentucky, and the Pueblo Chemical Depot in Pueblo, Colorado.

At BGAD, the chemical weapons filled with the nerve agents GB and VX pose particularly complicated destruction challenges that stem from a relatively diverse collection of weapons. The nerve agent-filled weapons at BGAD include about 52,000 115-mm rockets and 4,000 8-inch projectiles that are GB filled, and just under 18,000 115-mm rockets and 13,000 155-mm projectiles that are VX filled (NRC, 2005). Each rocket contains about 10.7 or 10.1 pounds of GB or VX, respectively. The 8-inch projectiles contain 14.4 pounds of GB, while the 155-mm projectiles contain 6 pounds of VX. The projectiles do not have energetics. The rockets, however, contain (in addition to agent) about 19 pounds of M28 propellant (nitroglycerine, nitrocellulose, and other additives) and about 3.2 pounds of Composition B explosive (59.5 percent RDX and 39.4 percent TNT) in bursters. There are also more than 15,000 155-mm projectiles filled with mustard agent, but the destruction of these items will utilize a static detonation chamber, and thus they are not a consideration of this study.

In response to a congressional mandate that destruction technologies other than incineration be identified and implemented at BGAD,[2] a process based on caustic hydrolysis of the agents has been adopted. The Blue Grass Chemical Agent Destruction Pilot Plant (BGCAPP) has been built at BGAD to destroy the BGAD stockpile using these technologies. The initial process used at BGCAPP to extract agent from the munitions varies depending on whether a rocket or a projectile is being destroyed. Figure 1-1 shows the processing flow of munitions through BGCAPP.

The following sections describe munitions processing at a high level. As will be discussed below, an agent washout step has been deleted from the munitions processing processes. The process flows are largely unchanged by the deletion of the washout step. Thus, the process descriptions below describe what will happen, with the divergences resulting from washout deletion called out where they occur.

PROCESSING OF ROCKET WARHEADS

Rockets will be processed in the rocket handling system (RHS), where the rocket cutting machine first separates the rocket motors from the agent-containing warheads. Uncontaminated motors will be sent off-site for destruction, and contaminated motors will eventually be sent to the energetics batch hydrolyzers (EBHs).[3] The rocket shear machine in the RHS punches and drains the warheads, generating two process streams, one liquid and one solid. The liquid stream consists of drained agents, while the solid stream consists of the drained warheads. Upon draining, the liquid agents will be directed through a three-way valve to a strainer, where solids that might clog the agent pump will be removed by filtration. The liquid will then be sent through a second discharge

---

[1] These sites were located in or near Anniston, Alabama; Pine Bluff, Arkansas; Newport, Indiana; Aberdeen Proving Ground, Maryland; Tooele, Utah; Umatilla, Washington; and Johnston Atoll in the Pacific Ocean.

[2] The National Defense Authorization Act for Fiscal Year 1993 (Public Law 102-484). See also, the Omnibus Consolidated Appropriations Act, 1997 (Public Law 104-208) and the Strom Thurmond National Defense Authorization Act for Fiscal Year 1999 (Public Law 105-261).

[3] The separated rocket motors will be accumulated in a box. They will then be monitored using headspace monitoring for agent contamination before leaving the agent controlled area.

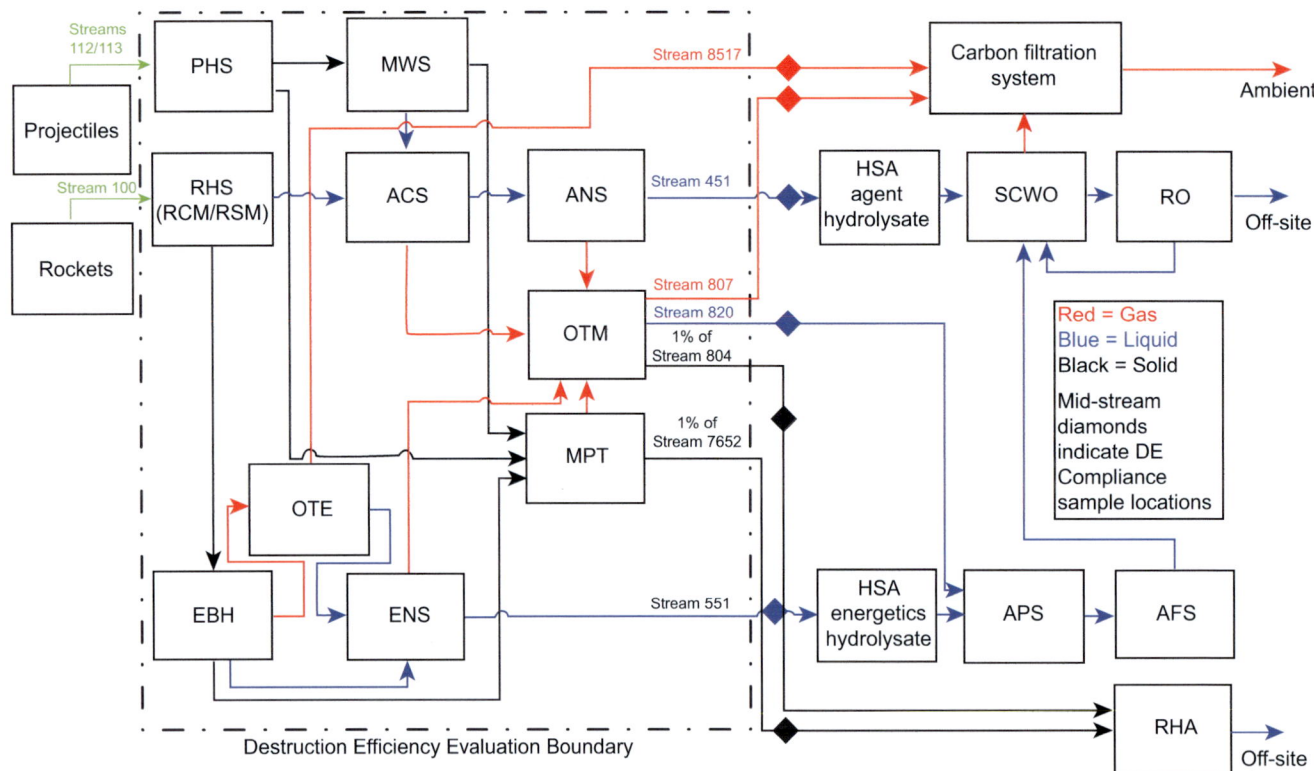

FIGURE 1-1 Block diagram showing the processing units and the flow of products at BGCAPP. Process streams represented by colored arrows represent projectiles or rockets entering, the process (green), solid materials (black), liquids (blue), and gases (red). Colored diamonds indicate measurement points along each process stream as they exit the munitions demilitarization building (i.e., the area where agent could conceivably be present). NOTE: Acronyms spelled out in front matter. SOURCE: Adapted from J. McArthur, environmental manager, BPBG, "Destruction Efficiency Considerations," presentation to the committee on September 9, 2015.

strainer that removes solids, which are primarily gelled or otherwise solidified agent that might interfere with subsequent hydrolysis reactions downstream. The filtered agents will then be transferred to the agent collection system (ACS).

In the original BGCAPP design, the rocket warhead bodies were to be washed out in the RHS with high-pressure, hot water that would have mechanically removed most of the residual agent on the interior walls of the munitions casing and broken up gelled and solidified materials that might have been present in the agents. This latter function, loosening and breaking up gels and solids, was deemed important because it had been estimated that solids and gels accounted for up to 20 percent of the GB in the munitions or up to 4 percent of the VX. The rinsate from the washout process was then to be directed through one of two strainer-pump-strainer trains via the three-way valve, in this case depositing the rinsate liquid into a tank that holds spent decontamination solution (SDS) (i.e., the SDS holding tank). The rinsate was then recombined with the drained agent before being sent to the agent neutralization system (ANS). The water washout ensured that nearly all agent would be processed through the ANS. This allowed measurement of any untreated agent at a single exit point from the Munitions Demilitarization Building (MDB), in order to meet the destruction efficiency (DE) requirements stipulated in the Research, Development, and Demonstration permit issued by the Kentucky Department for Environmental Protection to BGAD and BGCAPP on September 30, 2005, and in the Kentucky Administrative Regulations and the Kentucky Revised Statutes.

Before leaving the discussion of the generation of the liquid product in the RHS, it should be noted that, periodically, the filter media used in the strainers (referred to as filter socks) are expected to become clogged and require change-out. This process requires an operator equipped with appropriate personal protective equipment to enter the agent processing room to manually install new filter socks and place the clogged socks in a tray for eventual transfer to the EBHs for thermal destruction, as described later in this chapter. Manual entry is a time-consuming process with the potential to affect the overall safety and destruction schedule of BGCAPP.

The agent and wash water generated by the RHS would then be combined in the ANS to be hydrolyzed in a concentrated sodium hydroxide solution. The committee noted that when the process modified by elimination of the washout is used, only drained agent will be sent to the ANS. This hydrolysis occurring in the ANS eliminates the acute toxicity of the agent and generates a product that is referred to as hydrolysate, which is analyzed for residual agent. If the concentration of agent in the hydrolysate meets the specified release criteria (which are still to be determined), it is then transferred out of the MDB to the agent hydrolysate storage tank (located in the hydrolysate storage area [HSA]).[4] The hydrolysate is then destroyed in the supercritical water oxidation system. Destruction of the hydrolysate is important, because it eliminates the possibility of intentionally recovering the primary hydrolysis products, which could be reacted to regenerate the original agents. As such, hydrolysate destruction is a requirement of the CWC. Hydrolysate treatment is not part of the statement of task for this report. Assuming that the clearance criteria for effluents from the MDB remain the same, washout deletion should not impact hydrolysate treatment. Therefore, hydrolysate treatment is not discussed in this report. The headspace gases from the ACS and the agent neutralization reactor (ANR) have the potential to contain agent, and are, therefore, further processed through the off-gas treatment system (OTM).

The solid process stream from the RHS consists of the warhead cavities along with their energetics-filled bursters, which, under the original BGCAPP design, were to be washed out at the RHS to remove residual agent.[5] These solid components are then to be sheared into segments and transferred to the EBHs, where they are subjected to a hot (241°F at 1 atm), concentrated sodium hydroxide solution that destroys the energetics in the rocket warhead and dissolves a significant fraction of the aluminum.[6] There are to be three process streams from the EBHs: liquids, solids, and headspace gases.

The liquid effluent from the EBHs, which contains significant aluminum, is sent to the energetics neutralization reactors where it will be neutralized at 300°F and a pressure of 3.1 atm. This liquid product is fed to the aluminum precipitation system and then to the aluminum filtration system to remove the aluminum contained in this stream. The liquid effluent from the aluminum filtration system undergoes final treatment in the supercritical water oxidation, while the solid aluminum-bearing precipitate is transferred off-site for disposal.

The gaseous effluent from the EBHs will be sent to the energetics off-gas treatment system (OTE), which consists of a scrubber and a particle filtration system. The scrubber functions to remove ammonia and energetics degradation products that may be present in the off-gas, while the filter removes particles greater than 3 μm in diameter. The gaseous effluent was originally deemed to be largely free from the possibility of agent contamination because, due to washout, only small residual amounts of agent were to be sent to the EBH in the first place. This is now not the case. Gaseous effluent from the OTE flows into the MDB heating, ventilation, and air conditioning (HVAC) system.

The solid process stream from the EBHs consists of undissolved metal parts that are periodically removed and sent to the metal parts treater (MPT) for final treatment. The MPT heats these metal components to 1,000°F and operates under a superheated steam atmosphere (125 lbs/h) to ensure pyrolytic degradation of agent. These conditions have been deemed sufficient to destroy any residual agent.

The MPT has two product streams: scrap metal, which exits the MDB for off-site disposal, and off-gas, which is sent to the OTM. The OTM is distinct from the OTE in terms of gaseous input and unit operations, as described below.

The OTM receives gaseous waste streams from the MPT, SDS, ACS, and ANS, and also from the energetics neutralization system. The OTM consists of a thermal oxidizer (TOX), followed by a Venturi scrubber and a cyclone. The TOX functions to thermally oxidize any residual organics and to remove any particulate solids that might originate from the MPT. The OTM was not specifically designed to destroy agent but would likely do that effectively.[7] The TOX consists of two sections, an oxidizing section and a quench section. Sufficient air is added to the oxidizing section to ensure that the residual oxygen level in the TOX is at least 5 percent. Natural gas and a fine mist of water are added to maintain the operating temperature at 2,000°F. The TOX can operate at temperatures as high as 2,200°F if the process gas feed has a higher heating value. This higher operating temperature is needed when contaminated wood pallets, or shipping and firing tubes from the rocket handling system, are processed in the MPT. The oxidizing section has a minimum gas residence time of 2 seconds at a minimum temperature of 2,000°F. This residence time is required for destruction of polychlorinated biphenyls that will be present during processing of leakers. In the quench section, a fine mist of water is supplied to bring the exit temperature to 1,200°F. The MPT-OTM combination was designed to ensure that no residual agent would survive these combined processes.

The OTM has three process streams, all of which exit the MDB. The solid process stream (#7652) will be particulate

---

[4] Blue Grass Chemical Agent Destruction Pilot Plant (BGCAPP) Hazardous Waste Management Facility Permit, EPA ID KY8-213-820-105, issued on September 30, 2005, Appendix B, Compliance Schedule requires BGCAPP to submit to the Hazardous Waste Branch Manager, the Waste Analysis Plan, Target Release Levels.

[5] Neil Frenzl, resident engineering manager, Bechtel Parsons Blue Grass (BPBG), John Barton, chief scientist, BPBG, "Rocket Handling System/Munitions Washout System (RHS/MWS) Design Update," presentation on February 18, 2015, to the Committee on Chemical Demilitarization of the National Academies of Sciences, Engineering, and Medicine, Washington, D.C.

[6] John Barton, chief scientist, BPBG, "Update on High Temperature Destruction of Cyanide Activities," presentation on May 5, 2015, to the Committee on Chemical Demilitarization of the National Academies of Sciences, Engineering, and Medicine, Washington, D.C.

[7] Neil Frenzl, resident engineering manager, BPBG, John Barton, chief scientist, BPBG, "Rocket Handling System/Munitions Washout System (RHS/MWS) Design Update," presentation on February 18, 2015, to the Committee on Chemical Demilitarization of the National Academies of Sciences, Engineering, and Medicine, Washington, D.C.

from the TOX, which will likely be free of agent because the TOX operates at high temperature. The liquid process stream (#820) from the OTM, consisting of water used in the Venturi scrubber, will be sent to the aluminum precipitation system. The gaseous process stream (#807) will be transferred to the MDB HVAC system.

PROCESSING OF PROJECTILES

Because the projectiles do not contain energetics, their processing does not require the EBHs and associated downstream systems. Instead, the projectiles will be sent to the munitions washout system (MWS) via the projectile handling system. With the deletion of agent washout, the only part of the MWS still used is the cavity access machine. The cavity access machine hydraulically forces a ram into the projectile, collapsing the burster well into the agent cavity, allowing the agent to drain. The burster deformation is also a CWC requirement. The agent then drains by gravity and is fed via a three-way valve and strainers to the agent storage tank in the ACS. A washout nozzle array (nozzles providing high-pressure water jets) is integrated into the shaft of the burster well ram to provide a high-pressure water spray to clean and flush the cavity. This would have washed the projectiles out with high-pressure, hot water. The washout water would have been directed into the SDS tank. As in the case of the RHS, the agent and rinsate would then have been combined in the ANS, where they would have been neutralized. Now, with washout deletion, the washout nozzle is still present in the burster well ram, but it will not be used.

Without the water washout step, the projectile bodies will be sent to the MPT after the agent is drained. Originally, the projectiles would have contained only a small quantity of residual agent adhering to their surfaces due to agent washout. With the deletion of agent washout, more residual agent will be sent to the MPT. The residual agent in the projectiles will be destroyed by pyrolysis at high temperature (1,000°F) for 15 minutes. This produces decontaminated metal scrap and off-gas as byproducts. The MPT is designed for pyrolyzing the agent in projectile bodies and functions with trays that can accommodate up to 40 of the 155-mm projectiles drained of VX, or up to 18 of the 8-inch projectiles drained of GB. After thermal treatment, the projectile bodies are deemed free of chemical agent due to process knowledge and are released for disposal off-site. The MPT will also be used to treat the filter socks that capture solids during projectile agent drain operations. Gaseous effluent from the MPT flows to the OTM, where it undergoes treatment as described above for the rocket warheads.

CONCERNS RELATED TO THE OPERATION OF THE PLANT AS ORIGINALLY DESIGNED

The BGCAPP team identified a number of concerns regarding aspects of operation according to the original design,[8,9] which are summarized in the following six bulleted points:

- *The mixing of VX and water may result in an exothermic, autocatalytic reaction.* At the baseline plant at Tooele, Utah, a mixture of 3,000 pounds of VX and 1,200 pounds of water resulted in an exothermic reaction that elevated the temperature in the tank, causing emission of vapor and liquid carryover into a surge tank.[10] The observations were consistent with laboratory-scale studies conducted by Yang and co-workers, who concluded that attack by ethyl methylphosphonate on VX was an exothermic, autocatalytic process that would lead to elevated temperatures in large-scale systems (Yang et al., 1996). Frenzl and Barton suggested that the carbodiimide stabilizers may actually facilitate this reaction by deprotonating the ethyl methylphosphonic acid,[11] which is a chemical intermediate in the autocatalytic reaction pathway. Significant uncertainty remains in the understanding of the conditions that lead to the exothermic reaction. For example, it was noted that a water washout had been used to remove VX from ton containers at the Newport, Indiana, facility; yet this process did not result in a thermal runaway in the stored VX. But the wash water was not recombined with agent in that process. Weighing the totality of past experience, BGCAPP determined that the potential risk of a thermal runaway might be significantly reduced if the VX were not mixed with water.[12]
- *Water mixing with VX will result in agent gelling.* This is presumably due to aluminum salt formation with the hydrolysis product ethyl methylphosphonic acid. The aluminum originates from M55 rocket warheads.
- *The original strainers may have been undersized.* If this is true, then more demands would be placed on the strainer filtration function, a situation that could be exacerbated by the fact that the filterability of the gel is unknown.[13] It was estimated that if the GB muni-

---

[8] Ibid.

[9] Neil Frenzl, resident engineering manager, BPBG, John Barton, chief scientist, BPBG, John McArthur, environmental manager, BPBG "Changes to Munitions Drain Systems Update," presentation on May 5, 2015, to the Committee on Chemical Demilitarization of the National Academies of Sciences, Engineering, and Medicine, Washington, D.C.

[10] Neil Frenzl, resident engineering manager, BPBG, John Barton, chief scientist, BPBG, "Rocket Handling System/Munitions Washout System (RHS/MWS) Design Update," presentation on February 18, 2015, to the Committee on Chemical Demilitarization of the National Academies of Sciences, Engineering, and Medicine, Washington, D.C.

[11] Ibid.

[12] While the exothermic reaction is a concern in the storage tanks, they are not a concern in the neutralization reactors. The storage tanks are not designed to contain the reaction, in contrast to the neutralization reactors, which are designed to operate at high temperature.

[13] Ibid.

tions contain 20 percent solids, the originally forecast processing rates would likely not be achieved. This outcome would be a consequence of the requirement to change out filter socks when 50 percent full, which entails operators making a suited entry into the room to remove and replace the sock.

- *The RHS entrained air into the agent and washout water during agent washout.* First-of-a-kind testing showed that excessive air flow was required to control the water spray during washout and that the diaphragm drain pumps used to move the liquid into the ACS would not handle the air/agent or air/water feed.
- *Hydrogen is formed in the GB-filled rocket cavities.* Hydrogen gas that is entrained in the agent mixtures would be sent to the ACS along with air or nitrogen, which are used as purge gases to ensure that the hydrogen concentration would not exceed 25 percent of the lower flammability limit, thus limiting the potential for the hydrogen to accumulate in the ACS tank headspace. At the baseline incineration sites, the room air diluted the gas, which was then thermally processed with the agent; thus, there was no flammability danger. This dilution may not occur at BGCAPP, where the agent will be pumped into the ACS for a period of time until a sufficient quantity of agent has been accumulated for processing in the ANS.
- *Low points in the shared piping lengths would result in agent mixing with water, which would create the potential for accelerated corrosion.* In particular, hydrogen fluoride (HF or hydrofluoric acid) is formed during the GB campaigns, which can attack the carbon steel in these sections, but the rate of HF attack was initially anticipated to be slow. However, industry experience suggests that the rate of corrosion could be fast, and the presence of low points could create situations where the HF concentration is 1 percent or higher. Studies conducted by Army personnel show that corrosion approaches 500 mils per year (mil/year) when carbon steel was immersed in 1 percent HF.[14] It is worthwhile noting that at the baseline chemical demilitarization sites, carbon steel piping did not experience significant corrosion; however, a water washout was not used in these systems, an observation that provides further motivation for eliminating washout. If the GB mixing with water can be minimized or eliminated, the potential for high corrosion rates might be greatly reduced.

These concerns prompted a rigorous review that was conducted by a convened Technical Tiger Team. Their findings and recommendations, while not completely coincident with the concerns of the BGCAPP technical staff, were in agreement on multiple points. The Tiger Team generated several important recommendations that influenced the ultimate redesign:

i. Revise solids design criteria.[15] Originally, it was assumed that there would be up to 20 percent solids in all GB munitions, and up to 4 percent solids in the VX munitions. These estimates were felt to be conservatively high based on historical sampling evaluations. Accordingly, the criteria were adjusted.

For the GB-filled projectiles, a total of 5-10 percent of the GB fill is now projected to be un-drainable from 100 percent of the GB projectiles. There will be crystals in many if not most of the projectiles, but the GB-filled projectiles are expected to drain easily within the time allocated for projectile draining when crystals comprise less than 2 percent of the agent fill.

For the GB-filled rockets, it is estimated that 90-95 percent of the agent fill can be drained from 90 percent of the rockets within the time allocated. It is estimated that the remaining rockets will retain 30 percent of the GB-fill.

For the VX-filled rockets and projectiles, new criteria include the assumption that greater than 95 percent of the VX agent will drain from the agent cavity within the allotted drain time.

Generally, for projectiles containing crystals, only 2-3 percent of the drained agent is filterable at >500 μm. For projectiles that do not contain crystals and for all rockets, filterable content drops to 1-2 percent. This is significant because crystals >500 μm are capable of sequestering intact agent, resulting in incomplete treatment at that point in the process.

ii. Eliminate issues with air entrainment and hydrogen by deleting the water washout and adding a knockout pot with a dedicated vacuum pump.

iii. Eliminate issues with corrosion of the carbon steel in the RHS in the GB campaigns by eliminating the water washout.

iv. Eliminate heat buildup from the exothermic VX/water reaction by eliminating the water washout.

v. Reduce the strainer change-out frequency by increasing the size of the strainer equipment.

vi. Eliminate RHS low points to improve draining efficiency and mitigate corrosion.

---

[14] Ibid.

[15] Neil Frenzl, resident engineering manager, BPBG, John Barton, chief scientist, BPBG, John McArthur, environmental manager, BPBG "Changes to Munitions Drain Systems Update," presentation on May 5, 2015, to the Committee on Chemical Demilitarization of the National Academies of Sciences, Engineering, and Medicine, Washington, D.C.

WASHOUT DELETION MODIFICATIONS

The process at the BGCAPP was subsequently modified to eliminate the munitions water washout process, both from the MWS and the RHS. This change, which is included as part of a larger set of design changes that is designated Engineering Change Proposal-87 (ECP-87), brings planned BGCAPP operations into accord with the Tiger Team recommendations that advised elimination of the water washout process.[16,17] It is worthwhile noting that most of the operations remain the same. However, the process of extracting the agent from the munitions is being changed significantly, which will impact several downstream processes.

As a consequence of ECP-87, the agent drain systems were redesigned, with the objective of achieving several improved operational attributes. The specific changes are as follows:

- Accommodate two-phase flow from the RHS and MWS systems, making the system compatible with hydrogen from the RHS and adding the ability to remove gases from the liquid system;
- Modify the drain line and three-way valve to eliminate agent and water mixing;
- Meet design criteria for solids loading;
- Mitigate HF formation resulting from GB washout;
- Mitigate the heat generating, autocatalytic reaction of VX with water; and
- Control agent vapor generated by two-phase flow.

STATEMENT OF TASK

There may be unintended implementation impacts resulting from the deletion of the agent washout step in the MWS and the RHS units. These possibilities motivated Program Executive Office, Assembled Chemical Weapons Alternatives (PEO ACWA) to request that the National Academies of Science, Engineering, and Medicine convene an ad hoc study committee to examine the impacts of this design change on operations at BGCAPP. The committee was tasked with addressing the following issues:

- Assess the impact of the design change on plant operations and the impacts to plant throughput, taking into account revised rocket and projectile drain times, strainer change-out frequency, and metal parts treater throughput;
- Review and assess the calculations associated with the ability of the metal parts treater and thermal oxidizer to effectively process additional residual agent GB and VX contained in the drained rocket and projectile munition bodies;
- Review and assess the contractor's approach to the destruction efficiency (DE) calculations and provide any suggestions that support the DE confirmation process; and
- Assess the validity of process modeling conducted to date and recommend where additional modeling may be of benefit for understanding likely plant operation performance.

APPROACH TO THE STATEMENT OF TASK AND ORGANIZATION OF THE REPORT

Chapter 2 provides a more detailed evaluation of the effects of washout deletion on the individual unit operations, which can be compared to operations as they were initially conceived and designed. Chapter 3 addresses the impact of washout deletion on regulatory compliance and permitting—specifically, the ability of BGCAPP to demonstrate compliance with the 99.9999 percent ("six-nines") DE set forth in the Kentucky Revised Statutes. Chapter 4 discusses process modeling to date and what additional modeling may be useful.

REFERENCES

NRC (National Research Council). 2005. *Interim Design Assessment for the Blue Grass Chemical Agent Destruction Pilot Plant*. Washington, D.C.: The National Academies Press.

Yang, Y.C., L.L. Szafraniec, W.T. Beaudry, D.K. Rohrbaugh, L.R. Procell, and J.B. Samuel. 1996. Autocatalytic hydrolysis of V-type nerve agents. *The Journal of Organic Chemistry* 61(24):8407-8413.

---

[16] Neil Frenzl, resident engineering manager, BPBG, John Barton, chief scientist, BPBG, "Rocket Handling System/Munitions Washout System (RHS/MWS) Design Update," presentation on February 18, 2015, to the Committee on Chemical Demilitarization of the National Academies of Sciences, Engineering, and Medicine, Washington, D.C.

[17] There are other approaches to removing agent from munition cavities that do not involve water, and some of these were considered by the sponsor. However, these alternatives are not explored in this report because the washout deletion approach had already been selected and implemented by the time the committee was asked to conduct this study. Implementation has included both hardware and process changes, BGCAPP construction is now complete, and the plant is entering systemization. It would be impractical to make major changes to the plant at this late stage without a major driving impetus to do so. Consequently, the committee was tasked solely with investigating the impacts of washout deletion and providing recommendations to the sponsor on how to address these impacts.

2

Plant Process Changes as a Result of Washout Deletion

ROCKET AND PROJECTILE DRAIN OPERATIONS

Without a water washout, agent drain from munitions will be by gravity alone. The presence of gel and solids in the agent will limit the efficacy of the drainage process, so that the quantities of agent remaining in the munitions may not meet the design specification of 98 percent agent removal from the cavity of a given projectile or rocket warhead. The amount of agent remaining adhered to interior surfaces of rocket warheads and projectile bodies at the end of each drain cycle will depend directly on the agent's physical characteristics, which will in turn depend on the degree of agent degradation—gelling and crystallization—that has taken place within each munition. Gelation and crystal formation also affect the time required to drain a rocket or projectile of that portion of the agent that will flow by gravity.

Large numbers of munitions containing the nerve agents GB and VX have already been processed at the baseline sites; however, there are no data available on the rate or efficiency of unassisted drainage. The Blue Grass Chemical Agent Destruction Pilot Plant (BGCAPP) has information on the condition of GB and VX agent within a limited number of munitions from specific lots based on past sampling. The quantity of gel and crystals present in the GB munitions is highly variable among agent lots. Current estimates of the solid content are 5-10 percent for GB-filled munitions and 4 percent for VX-filled munitions.

Although data on agent in munitions are limited, a laboratory study of gel formation by the reaction of GB with metal halide salts found that gel and crystals form at lower temperatures and in higher quantities in munitions-grade GB than in Chemical Agent Standard Analytical Reference Material-grade GB (Yang, 2003). The extent of gel formation, and the nature of the gel product, also varied considerably depending on the metal salt examined, suggesting that varying contaminants within different batches of munitions could well explain the variable state of the agent. Yang (2003) found similar contaminants in some GB-filled munitions but does not report on the quantities or frequencies of occurrence of such contaminants. That study shows that even high-purity GB will form solids and viscous, oily products under conditions that could exist in storage. Therefore, these contaminants can be expected to be present in the less-pure munitions stored at the Blue Grass Army Depot, and there is the potential for degraded agent to be present in the munitions being processed.

Finding 2-1. Uncertainty in the number of munitions containing degraded agent and the degree of agent degradation is compounded by a lack of knowledge of the physico-chemical characteristics of degraded agent as they relate to drain times and amounts of residual agent retained in munitions at the end of the drain process. Better data are needed to properly estimate the time that will be required to process the nerve agent munitions through BGCAPP.

Recommendation 2-1. BGCAPP should gather data, such as mass drained, drain time, and any available information on physical state, for each individual munition during operations ramp up to assess the state of the agent fills and thus expected variability in drain times for each agent lot and type of munition. The acquisition of these data should continue throughout operations to continuously improve the quality of estimates as an aid toward planning of plant operations and to estimate completion times.

In addition to the possibility of problems being encountered in draining, the reconfigured agent transfer system has several additional components that may pose a risk of hazardous events during operations (Figure 2-1).[1] The agent is

---

[1] Neil Frenzl, resident engineering manager, Bechtel Parsons Blue Grass (BPBG), John Barton, chief scientist, BPBG, John McArthur, environmental manager, BPBG, "Changes to Munitions Drain Systems Update," presentation on May 5, 2015, to the Committee on Chemical Demilitarization of the National Academies of Sciences, Engineering, and Medicine, Washington, D.C.

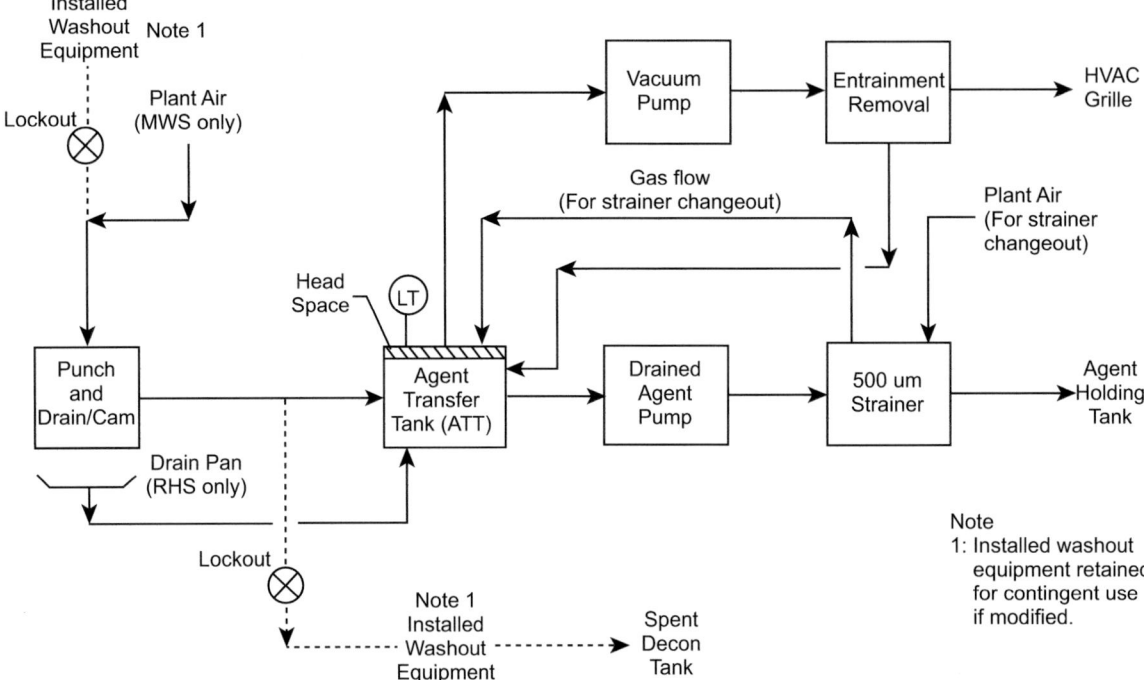

FIGURE 2-1. Agent transfer system block diagram. The agent transfer system is a subsystem of the PHS and RHS units, and takes the drained agent from the rocket warhead or projectile to the agent holding tank, where agent is gathered into batches for hydrolysis. NOTE: Acronyms are defined in the front matter. SOURCE: Neil Frenzl, resident engineering manager, Bechtel Parsons Blue Grass (BPBG), John Barton, chief scientist, BPBG, "Rocket Handling System/Munitions Washout System (RHS/MWS) Design Update," presentation on February 18, 2015, to the Committee on Chemical Demilitarization of the National Academies of Sciences, Engineering, and Medicine, Washington, D.C., used with permission of ACWA.

initially to be transferred into the agent transfer tank (ATT), which utilizes a vacuum pump to evacuate the ATT headspace and a drained agent pump to move the liquid agent to the strainers. BGCAPP technical staff members have rigorously specified requirements for the pumps; however, there will be the possibility of pump failures due to higher solids content in the liquid agent than originally planned. Excessive pressure in the drum demister (entrainment removal module), which is downstream of the ATT vacuum pump, is also a possibility. These failure possibilities will increase if high quantities of solids or gels drain from the munitions into the ATT. Any increase in failure events caused by these pump failures increases the probability of extending the agent processing campaign.

STRAINER SOCK LOADING

Processing munitions that have high gel or solids content is unavoidable. With the elimination of the washout step, the volume of solids being captured by the strainer system will increase. It may be that processing multiple high-gel or high-solids munitions will quickly overwhelm the strainer sock capacity, which would result in a higher frequency of change-out entries and a slowed processing rate. To accommodate this change in solids capture while minimizing the change-out frequency, the strainer unit size and the number of strainer baskets has been increased to reduce the number of worker entries required to change strainer socks. Even with increased filter capacity, the filter sock change-out frequency is estimated to be once every 3 days.[2] Moreover, it remains uncertain whether the knowledge of the solids load in agent fills is adequate to develop a design that will allow the planned change-out frequency to be achieved. The compatibility of this unit with the type of solids that could be encountered in the agent is not known. Because this unit is significantly larger than the original design, it can be anticipated that the volume of material to be removed during each individual change out will increase. The impact on the time needed for an individual change out due the larger unit volume needs to be determined. Data obtained during testing and continuing during plant operation can facilitate strategic planning to minimize the impact of high-solids munitions on processing rate. If filter socks have to be changed out much more frequently and/or take longer to change out than anticipated, this could have negative schedule consequences

[2] NRC Washout Deletion Committee Questions and Responses 151027 Set1, received via e-mail on November 6, 2015.

for BGCAPP operations, with implications for budget, treaty compliance, and reducing storage risk by the timely destruction of the stockpile.

Finding 2-2. Even with the change in filter sock capacity, the change-out frequency could become the rate-determining step in the processing of rockets and projectiles.

MPT AND PLANT OFF-GAS TREATMENT SYSTEM (OTM)

The agent destruction process is complicated by deletion of the washout step because agent treatment will now extend over several flow paths after drain operations. Residual agent remaining in rockets is to be sent to the energetics batch hydrolyzers (EBHs), while that remaining in projectiles is to be sent to the metal parts treater (MPT).

The MPT will be unaffected by the modified processing of the rocket components, which are transferred to the MPT only after immersion in hot caustic in the EBHs. No residual GB is expected to survive, and while VX droplets may survive the immersion step, the majority of these should be suspended in the hot caustic and not adhere to the undissolved rocket parts that are transferred to the MPT. Thus, the rocket components from the EBH are expected to be largely free of agent.

The projectiles entering the MPT will have more residual agent than the rocket warhead pieces entering the MPT because the projectile bodies are not processed through a hydrolysis step. Residual agent in the drained projectiles will vaporize and undergo pyrolysis in the MPT. The filter socks containing solids from drained projectiles will also eventually be sent to one of two MPTs for thermal decontamination and agent destruction.

The MPT is designed to process trays holding up to 40 of the 155-mm projectiles drained of VX, or up to 18 of the 8-inch projectiles drained of GB. BGCAPP has conducted numerical modeling of agent vaporization rates in the MPT under a simplifying set of assumptions. The simulation results suggest that, under scenarios involving fully loaded trays of projectiles containing 5 percent residual VX or 10 percent residual GB, the time to vaporize all agent from the projectiles only slightly exceeded the currently planned hold time in zone 1 of the two-zone MPT. Computational fluid dynamics modeling conducted by BGCAPP was used to estimate overall MPT processing times. It estimated that increasing the processing time from 60 minutes to 80 minutes for GB, or 90 minutes for VX, would ensure agent destruction. BGCAPP concluded that this increase would have minimal impacts on the schedule and duration of destruction campaigns. Computational fluid dynamics modeling also predicted that MPT processing could alternatively be conducted on partially filled trays of projectiles in order to reduce the quantity of agent processed in each batch. Either of these proposed strategies would ensure complete removal of residual agent from the metal parts.

Of particular interest regarding the deletion of the washout process is the added agent load to the OTM system where the residual agent that is vaporized and pyrolyzed in the MPT is sent prior to being forwarded to the Munitions Demilitarization Building (MDB) heating, ventilation, and air conditioning (HVAC) filter system. The OTM is responsible for processing gases from the agent collection system, the agent neutralization system, the energetics neutralization system, and the MPT. The deletion of the washout process may impact the OTM due to the potential presence of higher quantities of agent in the gaseous effluents from these units than originally planned for. If agent vaporization in the MPT is not complete, increasing steam injection could accelerate agent vaporization. This would result in an increased rate of agent pyrolysis products fed to the thermal oxidizer (TOX) process, but if these rates can be managed by the TOX, it would reduce the required residence time within the MPT. Additional steam may also suppress particle formation.

In spite of increased agent loading delivered to the TOX, BGCAPP estimates that the residence time in the TOX will be sufficient for complete destruction of any GB or VX that goes to the OTM from treatment systems.

In any case, with the elimination of the washout step, the volume of agent processed through the MPT and OTM will constitute a larger component of overall agent destruction efficiency as compared to the original configuration. With the increased number of agent processing paths, the methodology for determining the destruction efficiency will need to be adjusted. This is topic is discussed in Chapter 3.

Finding 2-3. Agent processed through the MPT and the off-gas treatment system will constitute a significant fraction of the agent destroyed at BGCAPP. This is a departure from the original design where almost the entire agent volume was being treated by hydrolysis.

Finding 2-4. Multiple mechanisms exist for controlling the MPT throughput rate to reduce instantaneous agent loading in the MPT and the off-gas treatment system. These include, but are not limited to, approaches such as increasing the residence time in zone 1 of the MPT, reducing the number of projectiles on each tray being processed and increasing the steam addition rate to the MPT.

Recommendation 2-2. BGCAPP should evaluate whether higher agent vaporization rates in the metal parts treater (MPT) can be accommodated by optimizing the operating parameters of the MPT, the off-gas treatment system, and associated systems.

EBH AND OFF-GAS TREATMENT SYSTEM FOR THE ENERGETICS NEUTRALIZATION SYSTEM (OTE)

As previously noted, the agent destruction process is complicated by the washout step deletion because the new processing conditions extend agent treatment over several flow process paths following drain operations. The impact of the change could impact the EBH and its off-gas treatment system (OTE).

Once the agent drain for rockets is completed, the rocket warhead is sheared; the metal parts are then sent to the EBHs where caustic is added. The mixture reaches a temperature of 241°F at a pressure of 1 atm to dissolve the energetics material in the rocket pieces. The resulting materials are sent in three directions. The caustic solution is sent to the energetic neutralization reactors (the main component of the energetics neutralization system) for final treatment of the energetics, the undissolved rocket parts are sent to the MPT, and the off-gases are sent to the OTE. In contrast to the OTM, the OTE does not contain a TOX. The OTE has a Venturi scrubber, which is meant to remove ammonia and energetics degradation products that may be present in the off-gas, and a particulate filter. While the OTE will also capture at least some agent, the effectiveness of the OTE for removing residual agent is not known and may not be adequate for agent destruction. Without the washout system in place, additional agent will remain bound to the rocket parts that are forwarded to the EBHs. Neither the volume of agent sent to the EBHs, nor that which is eventually forwarded through the EBH vapor space to the OTE, is known. Agent could find its way to the EBH vapor space by being volatilized in the EBHs rather than being hydrolyzed. In principle, the quantity of agent vapor should be modest because the boiling point of GB is 297°F (NIOSH, 2015a); however, it does have a significant vapor pressure. It is, therefore, conceivable that GB volatilization could be competitive with hydrolysis in the EBH. This is probably not an issue for VX, which has a much higher boiling point (568°F) and a much lower vapor pressure than GB (NIOSH, 2015b; Reutter, 1999). Fugitive agent making it through the OTE would then be vented to the MDB HVAC carbon filtration system.

Finding 2-5. With the deletion of munitions washout, some of the chemical agent from the rocket warheads will be sent to the EBHs. Some fraction of the agent introduced into the EBHs will be volatilized and then flow into the EBH OTE. The OTE does not have a TOX, so some of the agent transported from the EBH to the OTE may penetrate to the MDB HVAC.

Recommendation 2-3. BGCAPP should conduct modeling and experimental studies to bound the quantity of agent present in the OTE vent stream (stream #8517).

IMPACTS ON THE OVERALL MDB HVAC SYSTEM

The final area of potential impact from the elimination of the washout process is the overall MDB HVAC system. Key to this system is the activated carbon bed vapor scrubbing system, which is the last piece of process equipment through which the air passes after it exits the processing building and before it is sent into the atmosphere. The elimination of the washout step increases the quantity of agent contained in the projectiles and rocket warheads that can be partitioned into other parts of the building system. Some of this material may end up in the air within the room and may, therefore, be forwarded to the carbon beds for capture. In the case of VX, the assessment of the Tiger Team is that the effect will be negligible because the vapor pressure of VX is low.[3] This is not true for GB, on the other hand, and it may be that the change will increase agent load to the carbon filter banks. Therefore, there is the possibility for an increase in the carbon filter change-out frequency, along with an increase in the length of operations and disposal costs. The initial assessment of the Tiger Team was that at most only one additional change out of a carbon filter bank would be required; however, it is likely that there is uncertainty in this assessment. Figure 2-2 presents the results of modeling efforts of the expected GB venting load to the room air that provided the basis of these conclusions. One carbon filter change out is already planned.

The Tiger Team suggested that venting near the MDB HVAC would lower agent vapor levels, and, in fact, without implementing this approach, it estimated that 30-35 minutes would be required for GB concentrations to decrease to less than the short-term exposure limit (Figure 2-2).

Finding 2-6. During punch and drain operations, vapors are released directly to the room air and are exhausted through the MDB HVAC system. The primary mode of capture of these vapors is the carbon filter bank. This function is part of the original plant process; however, the washout deletion may affect agent concentrations in the gas phase that will be transferred to the MDB HVAC system.

Recommendation 2-4. BGCAPP should complete modeling to estimate the agent load to the carbon beds in the absence of a munition washout step to ensure that the lifetime of these beds is known.

REFERENCES

NIOSH (National Institute for Occupational Safety and Health). 2015a. *Sarin (GB): Nerve Agent*. Atlanta, Ga. http://www.cdc.gov/niosh/ershdb/emergencyresponsecard_29750001.html.

NIOSH. 2015b. *VX: Nerve Agent*. Atlanta, Ga. http://www.cdc.gov/niosh/ershdb/emergencyresponsecard_29750005.html.

Reutter, S. 1999. Hazards of chemical weapons release during war: New perspectives. *Environmental Health Perspectives* 107(12):985-989.

Yang, Y.-C. 2003. Synthesis and Characterization of GB Gels. Report 0151-090403.

---

[3] Neil Frenzl, resident engineering manager, BPBG, John Barton, chief scientist, BPBG, "Rocket Handling System/Munitions Washout System (RHS/MWS) Design Update," presentation on February 18, 2015, to the Committee on Chemical Demilitarization of the National Academies of Sciences, Engineering, and Medicine, Washington, D.C.

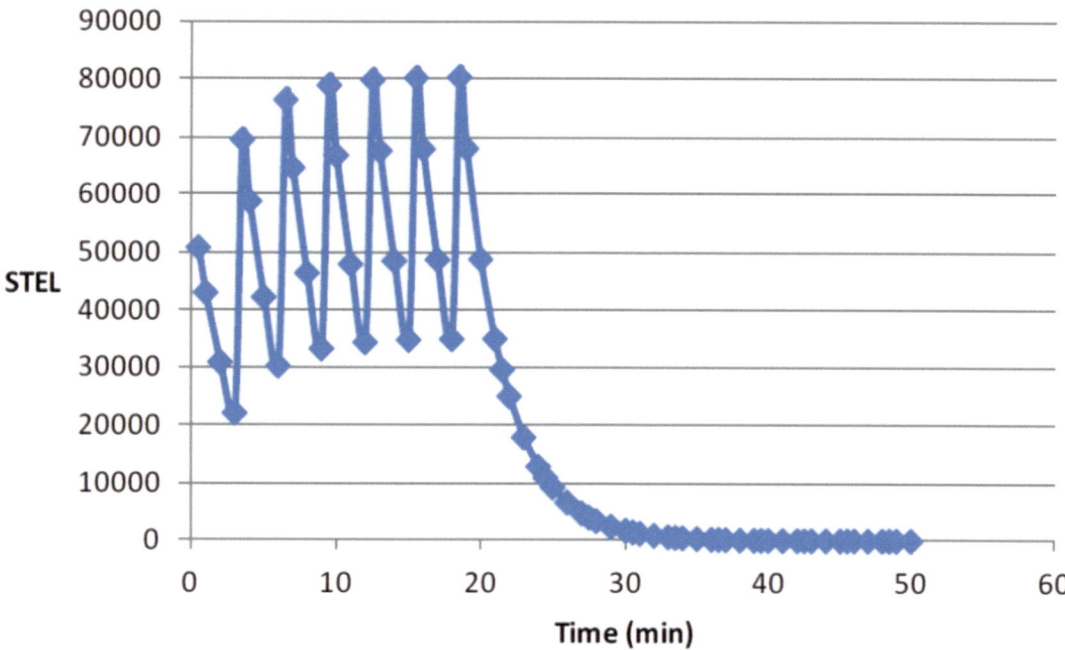

FIGURE 2-2 Worst-case GB scenario with venting to room. This scenario is based on a model in which seven punch and drain events are executed over 20 minutes, followed by an additional 30 minutes of room ventilation. The GB concentration, on the y-axis, is expressed in multiples of the STEL, which is the short-term exposure limit, and equivalent to 0.0001 mg/m^3. SOURCE: Neil Frenzl, resident engineering manager, Bechtel Parsons Blue Grass (BPBG), John Barton, chief scientist, BPBG, "Rocket Handling System/Munitions Washout System (RHS/MWS) Design Update," presentation on February 18, 2015, to the Committee on Chemical Demilitarization of the National Academies of Sciences, Engineering, and Medicine, Washington, D.C., and NIOSH (2015a).

3

Impacts on Calculation of Destruction Efficiency

CURRENT REGULATORY REQUIREMENTS FOR DESTRUCTION EFFICIENCY

The Kentucky Revised Statutes (KRS) 224.50-130 state that the Kentucky Energy and Environment Cabinet shall consider certain criteria in making a decision to issue a permit to any facility for the treatment or disposal of chemical agents. These criteria include whether the technology has been fully proven or demonstrated as effective to provide assurance of destruction or neutralization at an efficiency of 99.9999 percent (also known as "six-nines" destruction) for each compound to be treated. This statutory requirement is also reflected in the Kentucky Administrative Regulations (KAR) 401 KAR 34:350, where any proposed treatment or destruction technology for the treatment of nerve (i.e., GB and VX) and blister (i.e. mustard) agents must be proven in an operational facility of scale, configuration, and throughput comparable to the proposed facility for a period of time sufficient to provide assurance of 99.9999 percent destruction or neutralization (i.e., destruction efficiency, or DE) of each substance as determined by the following equation:

$$DE\ (\%) = (W_{In} - W_{Out} - W_{Res})/W_{In} \times 100\% \quad (1)$$

where

W_{In} = Mass feed rate of waste to the incinerator,
W_{Out} = Mass emission rate of the same waste present in exhaust emissions prior to release to the atmosphere, and
W_{Res} = Mass removal rate of waste via the incinerator residues.

This approach for assessing DE is directly applicable to the operation of an incinerator, where any surviving agent would necessarily reside in either the exhaust gases or in the solid residue. The Blue Grass Chemical Agent Destruction Pilot Plant (BGCAPP) will initially operate under the Resource Conservation and Recovery Act (RCRA) Research, Development and Demonstration (RD&D) permit EPA ID KY8-213-820-105, issued by the Kentucky Department for Environmental Protection (KDEP) to Blue Grass Army Depot (BGAD), BGCAPP, and Bechtel Parsons Blue Grass on September 30, 2005. Recognizing that BGCAPP is a neutralization facility, the KDEP modified the KAR DE calculation such that the current RD&D permit states that BGCAPP shall demonstrate 99.9999 percent DE on the initial batch of each chemical agent to be treated. According to RD&D Permit Condition T-9, DE is to be calculated as follows:

$$DE\ (\%) = 100 \times (M_1 - M_2)/M_1 \quad (2)$$

where

M_1 = Mass of agent per batch entering into the agent neutralization system (ANS) reactor, and
M_2 = Mass of agent per batch exiting the ANS reactor in the hydrolysate.

The current RD&D Permit, Appendix B Compliance Schedule, requires BGCAPP, at least 6 months before receiving waste, to submit to the Hazardous Waste Branch Manager, the agent neutralization reactor (ANR) DE Test Plan for the 99.9999 percent DE in the ANR (Paragraph 19), as well as a Waste Analysis Plan defining all target release levels as defined in the RD&D application and other areas yet to be determined (Paragraph 21). Although BGCAPP presented its original DE approach to the committee, it recognized that the deletion of the washout functions may impact this approach and presented two other options it is considering for the calculation of DE that could accommodate the change in the process flow pathways.

CURRENT APPROACH TO CALCULATING DESTRUCTION EFFICIENCY (APPROACH 1)

The equation for DE in the RD&D permit is a modified version of that found in the KAR and assumes that almost all agent would pass through and be treated in the ANS. However, a rigorous consideration of the process flow pathways shows that it was conceivable that agent could also be present as vapor generated by the draining and washout operations, and released into the processing room air, and from the ANS headspaces. Furthermore, agent adhering to drained projectiles would be transported to the metals parts treatment unit (MPT), and if any agent survived treatment there, it would then go on to the off-gas treatment system (OTM). However, with munitions washout, the reasonable expectation was that the quantity of agent in these alternative streams would be negligible with regard to calculating DE, and that almost all agent would pass through the ANS. Therefore, DE could be assessed merely by measuring the quantity of agent in the ANS after processing (M_2).

Individual projectiles will be weighed after draining to measure the completeness of the drain. This provides a reasonable estimate for the amount of agent transferred to the agent storage tank, which would then guide the quantity of caustic to be added to achieve the correct ratio for ensuring hydrolysis of the agent. Once the agent and caustic are combined, the de facto BGCAPP implementation of the RD&D permit to calculate DE would use agent *concentrations* in the feed and effluent streams; that is,

$$DE (\%) = 100 \times (C_1 - C_2)/C_1 \qquad (3)$$

where

C_1 = Concentration of agent per batch entering into the ANS reactor, and
C_2 = Concentration of agent per batch exiting the ANS reactor in the hydrolysate.

This approach greatly simplifies the calculation of the DE and the supporting analytical measurements. It was assumed that the concentration of GB in caustic entering the ANR would be 7.5 wt% agent, in accordance with the operating design specifications. This is equivalent to a fractional concentration of 0.075, the value for C_1 of Equation 3. Six-nines destruction requires that the concentration be reduced by a fraction equivalent to $(0.999999) \times (0.075) = 0.074999925$, which means that maximum allowable residual concentration would be $(0.075) - (0.074999925) = 0.000000075$. This value, more commonly expressed as 75 parts per billion (ppb), is the maximum allowable concentration in the ANS before exiting as hydrolysate effluent (stream #451, the C_2 value of the Equation 3). Analytical measurement to ascertain destruction to this level was based on measuring a concentration of 75 ppb or less.

This concentration-based DE assumes that the volume entering the ANS would be equivalent to the volume exiting. However, the concentration-based DE approach will be difficult or impossible to implement because, after washout deletion, the agent will now be partitioned between multiple processing pathways.

IMPACT OF WASHOUT DELETION ON THE CALCULATION OF DESTRUCTION EFFICIENCY

The committee recognizes that the following discussion is complex in places. This is the nature of the system being discussed. The reader is directed to Figure 3-1 for help in following the discussion. Under the current permit, the expectation is that all but a trace of agent will pass through the ANS, which includes the ANR. The DE would be calculated, as described above, by assuming a GB recipe of 7.5 wt% agent in caustic entering the ANR and would require an analytical clearance of less than 75 ppb in the exiting hydrolysate effluent to demonstrate 99.9999 percent destruction. The current permit does not address VX treatment. The VX treatment campaign will be conducted under a RCRA Part B permit that is currently being prepared for submission to KDEP. However, under the current plant configuration, the expectation is that the DE would be calculated by assuming a VX recipe of 16.6 wt% agent in caustic entering the ANR with an analytical clearance of less than 166 ppb in the hydrolysate effluent to demonstrate 99.9999 percent destruction. The comparison of the percentage concentration of the agent(s) in the ANR with the fractional concentrations in the hydrolysate for ascertaining DE is referred to by BGCAPP as DE calculation Approach 1.[1]

Upon washout deletion, however, a greater amount of residual agent will now be treated within the energetics batch hydrolyzers (EBHs), the energetics neutralization system (ENS), and the MPT. Therefore, the DE calculation established in the current RD&D permit would no longer account for all agent treatment effluent or residue. Under the new configuration, agent would be distributed among multiple solid, liquid, and vapor process streams. The mass of agent entering the ANS, EBHs, and MPT will not be known and will vary on a batch-to-batch basis. In addition, it is not certain that the proposed agent recipe for the ANS would be applicable to agent destruction in the ENS to affect an agent concentration in the neutralized energetics hydrolysate less than the assumed clearance concentrations of 75 (or 166) ppb in Approach 1. Consequently, the assumed clearance concentrations presented by BGCAPP technical staff are now less valid than before and cannot be used for calculating a DE value that would be in accord with the KRS, the KAR or the RD&D permit. Therefore, it will be necessary to modify

---

[1] John McArthur, environmental manager, Bechtel Parsons Blue Grass (BPBG), "Destruction Efficiency Considerations," presentation to the committee on September 9, 2015.

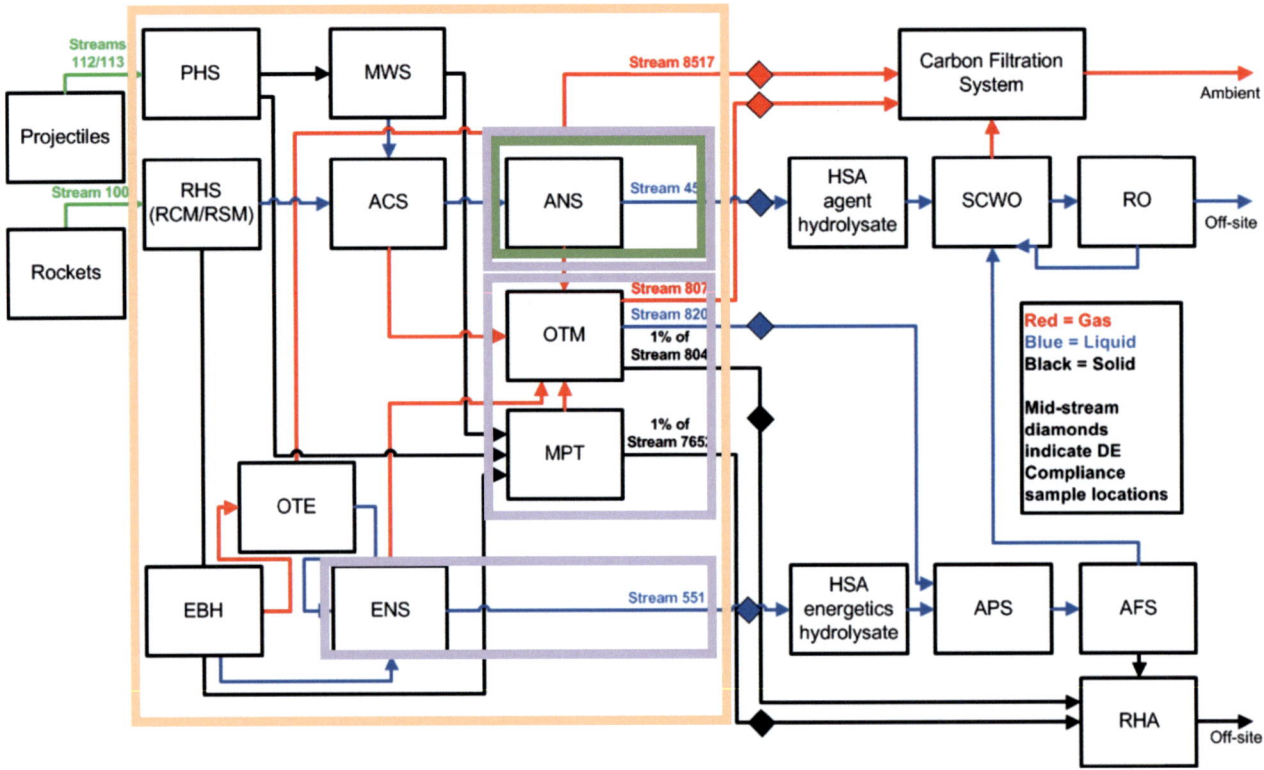

FIGURE 3-1 Process flow diagram for destruction efficiency calculation under Approaches 1, 2, and 3. Approach 1 is reflected by the green box around the ANS. Approach 2 is reflected by the purple boxes around the ANS, MPT/OTM, and ENS. Approach 3 is reflected by the large orange box. NOTE: Acronyms are defined in the front matter. SOURCE: Adapted from John McArthur, environmental manager, Bechtel Parsons Blue Grass, "Destruction Efficiency Considerations," presentation to the committee on September 9, 2015.

the existing permit to establish an alternate DE calculation methodology, staying within the statutory and regulatory DE requirements for 99.9999 percent destruction or neutralization as shown in the Equation 1.

The partitioning of agent across additional streams also introduces the need to conduct more monitoring than in the original DE approach. The agent concentrations in some of these streams could be quite low. This could pose a challenge in identifying monitoring technologies or strategies of sufficient sensitivity to measure these concentrations, be they new technologies introduced into BGCAPP or the adaptation of existing BGCAPP monitoring technologies. Additionally, the validation and acceptance of new monitoring technologies or strategies takes time and could have a schedule impact on BGCAPP operations. This committee is not in the position to identify the magnitude of these challenges and recommend solutions, but believes this is a potentially significant issue that BGCAPP management needs to be aware of.

Finding 3-1. With washout deletion, the current methodology for calculating whether BGCAPP meets the statutory and regulatory requirement for a DE of 99.9999 percent will no longer be applicable. This is because the drained and washed out agent will no longer go only to the ANS. Instead, a greater amount of residual agent remaining in the rocket warhead pieces will now be processed in the EBHs and the ENS, and residual agent in the projectiles and possibly on rocket warhead pieces will now be processed through the MPT.

Finding 3-2. The partitioning of agent across additional process streams will introduce monitoring challenges that BGCAPP needs to be aware of. It may be difficult to identify monitoring technologies or strategies of sufficient sensitivity to measure what are expected to be low concentrations of agent in some streams. Additionally, the identification of new technologies or strategies carries the risk of a negative impact on the BGCAPP schedule.

The original approach to be used in the DE calculations will no longer be applicable after washout deletion. Without the water washout, more agent will remain in the projectile bodies and rocket warheads, and any new DE determination will need to address issues raised with the new configuration. More agent from projectiles exiting the projectile handling system (PHS) will be partitioned between the ANS and the MPT, and more agent from rocket warheads will be partitioned between the ANS and the EBH/ENS. Furthermore, the fractions of agent that will be partitioned into the MPT or into the EBH/ENS will vary depending on the percentage of undrained agent remaining in each munition. Because the MPT, EBH, and ENS units are operated at high temperature and the energetics hydrolysis systems use a high concentration of sodium hydroxide, it may be that the agent partitioned into these units will be completely destroyed. However, the efficiency of these units for destroying agent present at the higher loadings after washout deletion is not known. For example, the EBHs were designed to neutralize energetics, and it is possible that a fraction of the agent may survive the EBHs.[2] Therefore, the effluents from the EBH/ENS must be considered as part of the overall DE calculation.

Agent Partitioning to the EBH and ENS Units

There is a possibility that agent will be present in the effluents from the EBHs. Any such agent will need to be considered as part of the overall DE calculation. Prior experience with the caustic hydrolysis of GB and VX suggests that the agents will be completely destroyed by caustic present in the EBHs. This expectation is consistent with calculations by BGCAPP that predict that there will be sufficient excess caustic to ensure quantitative destruction, and with the fact that GB is soluble in the caustic solution. However, VX is not soluble in the caustic solution and is highly surface adsorptive. Hence, it has a better chance of surviving as a result of either incomplete mixing or sequestration in crevices and pores in the metal parts. Further, there is also a chance that a fraction of the agent, particularly GB, which has a lower boiling point than VX, will be partitioned into the EBH off-gas treatment system (OTE). These considerations support the conclusion that agent partitioning into the EBH and ENR effluents must be accounted for in the DE calculation, which will consequently be complicated by the fact that there are three effluent streams that must be accounted for. These are considered in turn in the following text.

Undissolved metal parts will be removed from the bottom of the EBHs and transferred to the MPT. Based on previous experience in the chemical demilitarization program and process modeling, the high temperature and long residence time in the MPT is expected to destroy any agent remaining on the metal parts.[3] Therefore, agent on metal parts from the EBHs processed in the MPT is not expected to affect calculation of the DE; that is, measurement of process stream #7652 for agent would not be required.[4] However, the MPT was not initially intended for treating larger quantities of agent, as will be the case with the deletion of washout, and there is a possibility that volatilized agent could enter the OTM. Due to the change of circumstances, it would be necessary to show that additional agent loading to the MPT would not affect the calculation of DE through either calculation or demonstration; this is considered in more detail below.

Finding 3-3. It is likely that the combination of the EBH and MPT treatment conditions will be sufficient to destroy all residual agent residing on rocket warhead pieces. This, however, needs to be shown through either calculations or demonstration.

Recommendation 3-1. BGCAPP should calculate or otherwise demonstrate a 99.9999 percent ("six-nines") destruction efficiency (DE) for residual agent residing on rocket warhead pieces exiting the metals parts treatment unit (MPT). This would provide assurance that the solid effluent from the MPT (stream #7652) generated during rocket campaigns is free of agent to ensure compliance with DE requirements.

The second effluent stream from the EBHs is liquid hydrolysate, which will be transferred to the ENS, where the hydrolysate will be further treated with caustic but at higher temperature and pressure (300°F and 3.1 atm). The ENS generates a liquid effluent stream that is transferred outside the agent-controlled area to the hydrolysate storage area energetics hydrolysate storage tank (process stream #551). BGCAPP calculations suggest that there is sufficient caustic in the ENS to completely destroy any agent surviving the EBHs. This expectation is further supported by the higher temperature and pressure used in the ENS. Nevertheless, the committee is not aware of any evidence that the system will achieve satisfactory DE criteria. To ensure proper accounting of agent destruction, it would be necessary to determine the residual agent levels in this stream.

Finding 3-4. Complete destruction of augmented agent loadings passing through the EBH/ENS system has not been demonstrated.

Recommendation 3-2. BGCAPP should demonstrate satisfactory destruction efficiency for agent serially treated with caustic under the same conditions as those present in the energetics batch hydrolyzers (EBHs) and the energetics neutralization system at agent loadings equivalent to the highest quantities anticipated to be treated by the EBHs without washout.

---

[2] Resource Conservation and Recovery Act (RCRA) Research, Development and Demonstration (RD&D) Revision 5 Submission, 24915-000-GPE-CGPT-00001, filed with the Kentucky Department for Environmental Protection (KDEP) on February 20, 2014.

[3] The metal parts treater treats its process streams at 1,000°F for at least 15 minutes.

[4] The RD&D Revision 5 Submission states, "[A]fter demonstrating 99.9999 percent DE [destruction efficiency] for agent hydrolysate, validated process controls and statistical testing may be used in lieu of analyzing all batches of agent hydrolysate" Section 3.2.2, p. 66. It is assumed this concept would apply to each of the DE calculation approaches and measurement would not be needed after the original validation for any waste stream included within the DE calculation.

The third effluent stream from the EBHs is the headspace gas. This stream has the potential to be problematic because it is conceivable that some fraction of the agent will volatilize when it hits the hot caustic of the EBHs instead of being hydrolyzed. This is more of a concern for GB, because it has a higher vapor pressure than VX. If a fraction of the agent were vaporized, it would be transferred to the EBH off-gas treatment system (OTE), instead of being transferred to the ENS.[5] The OTE system consists of a scrubber that is designed to remove acidic gases. Removal of other organic compounds will also occur here, but the efficiency will be dependent on their solubility in water, since the scrubber uses aerosolized water droplets to capture contaminants. The water solubility of VX is limited, which suggests that the efficiency of its removal by the OTE may not be high; conversely, GB is water soluble, which suggests a higher removal efficiency for GB in the scrubber. However, the OTE's agent removal efficiency is not known for either agent. This is important because off-gas from the OTE (stream #8517)[6] could contain measurable agent, which would not undergo any further filtration or thermal oxidation before being sent to the carbon filter banks in the Munitions Demilitarization Building (MDB) heating, ventilation, and air conditioning (HVAC) system.

It is worthwhile noting at this point that, as discussed in Chapter 2, the load to the MDB HVAC system has always included agent vaporized into the room air during munition drain processes.[7] This is still intended, and committee discussions considered implementation of local, shrouded ventilation around each munition as it is punched and drained to reduce the buildup of agent vapor in the rocket handling system (RHS) and munitions washout system (MWS) rooms, which would more efficiently conduct vapor to the MDB HVAC. The MDB HVAC consists of both the activated carbon filter beds, and the ducting that transports air from different rooms to the carbon beds. The BGCAPP RD&D Permit Revision 5 Submission describes the MDB HVAC system function as maintaining a negative pressure environment in the MDB and a flow of air from areas of low contamination probability to areas of higher contamination probability; these pressure and flow characteristics serve to remove agent from the air prior to discharge to the atmosphere after the air stream has passed through other air pollution control systems, including the OTM and OTE.[8] The RD&D Permit Revision 5 Submission, however, also anticipates that that MDB HVAC system controls contaminants that might be released from the process as a point source or as a fugitive emission.[9]

The fact that the MDB HVAC may receive agent-containing vapor is also recognized by the existing RD&D Permit, in Permit Condition T-11, in that it requires monitoring of the MDB HVAC effluent to ensure no confirmed detectable agent emissions. MDB HVAC effluent monitoring for agent is also required by the Title V Air Quality Permit,[10] issued to BGAD on June 6, 2011, for the BGCAPP, which requires that the BGCAPP emissions not exceed the General Population Limits specified by the Centers for Disease Control and Prevention for Lethal Nerve Agent VX and Lethal Nerve Agent GB (6×10^{-7} and 1×10^{-6} mg/m^3 respectively) at the BGAD property boundary. In addition, the RD&D Permit Revision 5 Submission anticipates that some agent-contaminated carbon will be generated, in that it provides for the off-site disposal of agent-contaminated carbon from the MDB HVAC filters.[11,12] Some of the agent will also likely have deposited onto the surface of the HVAC ducting leading to the carbon filtration beds, representing yet another reservoir into which agent is partitioned. However, this is expected to be on the order of a single molecular layer and, hence, would account for at most a small fraction of the total agent, certainly much less than one part in a million corresponding to the upper limit permissible in achieving 0.999999 DE. Hence, this stream is not further considered in assessing DE.

The fact that agent-bearing vapor from the munitions drain operations are directed to the MDB HVAC suggests that the OTE gaseous effluent could be handled in the same way—that is, sent to the MDB HVAC—and that all agent in the stream not be counted in the DE calculation. Note that this is how the stream was to be handled under the original configuration (with the water washout), because it was not anticipated that a gas waste steam from the EBH entering the OTE and going directly to the MDB HVAC system could contain appreciable quantities of agent. However, increases in the agent loads resulting from washout deletion could make approval of any application for a permit revision problematic.

If BGCAPP could count agent trapped on the carbon filter banks of the MDB HVAC system as destroyed, then agent partitioned into the OTE gaseous effluent stream #8517 would not affect whether BGCAPP achieves DE criteria, because the multiple banks of both particulate (HEPA) filters and activated carbon filtration banks that comprise the MDB HVAC system will capture all of the agent exiting the MDB. However, BGCAPP believes that it would

---

[5] Note that the name of this unit may be confusing in that the EBH off-gas treatment system (OTE) does not receive vapor from the energetics neutralization system (ENS), only the energetics batch hydrolyzer (EBH).

[6] There are two effluent streams from the OTE, but the scrubber water, which is sent to the ENS, should not create an issue in that the high caustic in the EBH is expected to hydrolyze agent within this source.

[7] E-mail from Kyle Conway, BGCAPP, to Jim Myska, committee study director, on December 8, 2015.

[8] RCRA RD&D Revision 5 Submission, 24915-000-GPE-CGPT-00001, filed with KDEP on February 20, 2014, p. 41.

[9] Ibid, p. 42.

[10] Ibid, pp. 66 and 42.

[11] Ibid, p. 74.

[12] The committee was not tasked with evaluating whether other BGCAPP environmental permits and documentation (e.g., the National Environmental Policy Act) would need to be amended to accommodate the internal operational changes resulting from washout deletion.

not be allowed to take credit for removal of agent vapor by carbon bank adsorption prior to release of exhaust to the atmosphere because the DE does not allow for removal, only destruction.[13] Consequently, the mass of agent in the OTE effluent stream will require measurement for inclusion in the DE calculation.

Finding 3-5. An unknown fraction of agent entering the EBHs during the rocket campaigns may undergo volatilization instead of hydrolysis. Volatilized agent will be processed through the OTE system (stream #8517), which is not designed to destroy agent. Agent escaping the OTE will be removed to the MDB HVAC carbon filter banks, together with fugitive agent emissions from the munition drain processes. Because agent partitioned into these pathways cannot be counted as destroyed, and because BGCAPP believes it is not likely to be allowed to take credit for removal of agent vapor by carbon bank adsorption in the MDB HVAC prior to release of exhaust to the atmosphere, the implementation of washout deletion will require significant permit modifications and has the potential to prevent BGCAPP from achieving DE criteria.

Recommendation 3-3. For all of the gaseous process streams, BGCAPP should rigorously demonstrate that negligible agent is partitioned into the munitions demilitarization building (MDB) heating, ventilation, and air conditioning (HVAC) carbon filter banks under all conditions that could arise during the rocket campaign. BGCAPP should provide for monitoring of the OTE effluent stream (#8517) with analytical sensitivity sufficient to ensure that destruction efficiency criteria are achieved before they enter the MDB HVAC system.

One possible way to address the possible agent vapor would be to process OTE effluent through the OTM, as is planned for the air flows from the agent collection system (ACS) and ENS. The feasibility of this approach would be dependent on the impact on the thermal oxidizer (TOX) of this additional gaseous influent flow. Specifically, the addition of the OTM gaseous effluent to the TOX would increase the total flow rate and reduce the gas residence times in the TOX. However, results from BCAPP modeling of higher residual agent levels retained in projectiles treated in the MPT due to washout deletion estimated that TOX residence times would only decrease to 4.3 seconds for VX and to 2.1 seconds for GB.[14] These residence times are substantially longer than the minimum 0.5 second residence time required to ensure agent destruction; for the leaker campaign, the minimum TOX residence time increases to 2 seconds to ensure destruction of polychlorinated biphenyls.[15] Because the residence times predicted for the agent are longer than even the very conservative time used for polychlorinated biphenyls, they are highly likely to be sufficient to destroy additional agent. Therefore, based on these BCAPP estimates, there should be sufficient additional TOX capacity available to accommodate additional gaseous influent streams from the OTE effluent. Another possibility would be to add a TOX to the OTE. However, BGCAPP construction has been completed and the plant is entering systemization. The procurement and installation of another major piece of equipment would cause schedule delays. Also, the BGCAPP footprint is small and is already tightly packed with equipment. This could make the installation of an additional piece of equipment problematic. For these reasons, the committee believes that routing OTE effluent through the OTM is preferable to adding a new TOX to the OTE.

Recommendation 3-4. BGCAPP should examine the possibility of routing the gaseous effluent from the OTE (energetics batch hydrolyzer off-gas treatment system) into the OTM (off-gas treatment system). This would eliminate the biggest uncertainties in M_{Out} exiting the munitions demilitarization building (MDB), because it is likely that any agent surviving the OTE would be destroyed in the OTM. The number of gaseous streams from processing units exiting the MDB would be reduced to a single stream—namely, the off-gas from the OTM—and would be less likely to contain significant agent as a result of off-gas passing through the thermal oxidizer.

Before leaving the discussion of agent partitioning into the EBH/ENS units, one additional gaseous effluent stream needs to be considered. The ENS also generates headspace gas that is sent to the OTM, where it is passed through the TOX, the Venturi scrubber, and cyclone. It is very likely that residual agent from the ENS headspace gas would be destroyed in the TOX, and hence, this is unlikely to significantly contribute to the amount of agent in the OTM emissions to the MDB HVAC system (stream #807).

Agent Partitioning to the MPT and OTM Units

As noted above, the MPT receives drained projectiles from the PHS and pieces of rocket warheads from the EBHs. It is likely that the majority of residual agent on rocket warheads will have been destroyed in the EBHs, although a fraction might survive and be sent to the MPT. With deletion of the washout step, more agent will be sent to the MPT with the drained projectile bodies. The MPT is also to be used to pyrolyze agent in the strainer socks from munitions drain operations, and because larger strainer socks are now planned, this will also be the source of a larger agent load

---

[13] John McArthur, environmental manager, BPBG, "Destruction Efficiency Considerations," presentation to the committee on September 9, 2015.

[14] George Lucier, deputy chief scientist, BPBG, "Impacts of Washout Deletion on Metal Parts Treatment and Thermal Oxidizer," presentation to the committee on September 9, 2015.

[15] Ibid.

entering the MPT. Therefore, the MPT will now treat more residual agent than planned under the original operating configuration.

While the committee's initial considerations indicate that the MPT should be able to destroy the additional agent from this source, the fact that the MPT will have to handle more agent will compel analysis of the effluent streams emanating from the MPT, of which there are two. The first is the solid projectile bodies and pieces of rocket warheads treated by the MPT that comprise stream #7652; the treated solids are considered to be agent-free as a consequence of the high MPT treatment temperature and residence times noted above.

The second stream is the gaseous MPT effluent, which is subjected to additional treatment through the OTM. The OTM consists of a TOX, the Venturi scrubber, and a cyclone. Under the original configuration, with munition washout, the OTM received gaseous waste streams from the ACS/toxic storage tank, the ANS, and the ENS, in addition to the gas from the MPT. However, as discussed above, under the new configuration, without washout, the ENS and the MPT will see more agent than originally planned, and, therefore, so will the OTM. Because the overall quantity of agent that now will have to be treated by the OTM is not known, and because the OTM was not specifically designed to destroy agent to the six-nines DE criteria, the three effluent streams from the OTM will need to be measured for agent. To rigorously evaluate the DE, the potential agent contained in the three effluents (gaseous, liquid, and solid) from the OTM would need to be included in the calculation of the total agent effluent from BGCAPP.

The amount of agent present in the gaseous effluent emanating from the OTM is likely to be very low, on account of the very high temperatures in the TOX. Hence, the residual agent in the headspace gases emanating from the ACS, ANS, ENS, and MPT could, possibly, be completely destroyed. If this assessment is correct, then effluent from the OTM would not convey any agent to the MDB HVAC system via stream #807, but this cannot be assumed a priori. However, it is worthwhile noting that pilot or experimental evaluations of the efficacy of the OTM for handling higher quantities of intact agent have not been conducted. Any measureable agent in OTM effluent stream #807 will need to be included in the DE calculation.

The second effluent stream from the OTM is water from the Venturi scrubber that is sent to the aluminum precipitation system (denoted as stream #820). Because this is downstream of the TOX, it is not likely to contain a significant quantity of agent; however, this has not been demonstrated. Calculation of the DE would require measurement of the flow rates of this stream and the concentrations of agent within it. The resulting agent mass flow rates would then need to be included in the overall DE.

Finding 3-6. It is likely that the combination of the MPT and the OTM will completely destroy any agent entering the MPT. However, after washout deletion, the OTM will receive gaseous streams from other sources that may contain more agent than originally planned. It is currently unknown whether the OTM can adequately treat the combined load of all streams after washout deletion.

Recommendation 3-5. BGCAPP should measure solid, gaseous, and liquid effluents from the OTM (off-gas treatment system) during initial projectile campaigns to ensure that these effluents meet the destruction efficiency criteria.

The third effluent stream emanating from the OTM consists of solid waste (stream #804) that will likely consist of particles formed in the TOX, which must periodically be removed. As in the case of the metal scrap from the MPT (stream #7652), this material will have been generated by a very-high-temperature process and, therefore, is very unlikely to contain agent. Stream #804 will thus be handled through the residue handling areas for off-site shipment. This conclusion would have to be documented to the satisfaction of the KDEP through validated process controls, as set forth in the BGCAPP RD&D permit (see footnote 5), and statistical testing. Measurement of agent in this stream for calculation of the DE may be deemed unnecessary.

Finding 3-7. The solid waste stream from the OTM should be agent-free. This conclusion will need to be demonstrated to the Kentucky Department for Environmental Protection based on validated process controls and statistical testing.

ALTERNATIVE APPROACHES TO CALCULATING DESTRUCTION EFFICIENCY (APPROACHES 2 AND 3)

BGCAPP is working with KDEP to identify appropriate methods to calculate DE after washout deletion and is considering two alternative approaches for generating valid DE calculations with their attendant measurements. These are referred to as Approach 2 and Approach 3 (Approach 1, discussed above, represented the original plant operation that assumed that nearly all agent would be processed through the ANS and will no longer be applicable after washout deletion). All three approaches are presented graphically in Figure 3-1. In short,

- Approach 2 would evaluate DE by measuring agent mass in the feed and effluent streams in the individual treatment units within BGCAPP, except for the OTE (Figure 3-1, individual units to be measured outlined in the purple boxes).
- Approach 3 would evaluate DE by estimating the mass entering the MDB in the individual munitions or batch or munitions, and by measuring the mass in each waste stream as it leaves the MDB (Figure 3-1, orange box).

Calculation of Destruction Efficiency Under Approach 2

Under Approach 2, it is assumed that all agent will be partitioned among three processing units: the ANS (as was originally conceived), the MPT/OTM (considered as a single processing unit), and the ENS. To calculate the DE value as presented in Equation 2, as stipulated by the current RD&D permit, the quantities of agent in the process streams entering (M_{In}) and exiting (M_{Out}) these processing units must be measured or known.[16] By summing M_{In} values and M_{Out} values for each of the three units, DE could be calculated using a modified version of Equation 2. However, the problem with this approach is that there are no reasonable means for measuring all of the M_{In} or M_{Out} values.

There are a total of three liquid streams, two gaseous streams, and three solid streams that contribute agent to these processing units (M_{In}). Liquid effluents from the ACS (entering the ANS), and from the OTE and EBHs (both entering the ENS) would need to be measured for agent concentration and volume to enable calculation of agent mass entering these processing units. Similarly, agent concentrations and volumes would be needed for the gaseous streams entering the OTM from both the ACS and the ENS. Finally, agent mass would need to be measured on projectile bodies entering the MPT from the MWS and the rocket pieces from EBHs, and on filter socks from munitions drain operations. Summing the agent masses from these streams on a per-munition or per-batch basis would provide a total M_{In} value. However, there are currently no analytical devices in place to accomplish the needed measurements anywhere on the feed side of these units, so determining M_{In} values would require additional measurement methodologies.

Multiple measurements would also be needed for calculating total agent exiting the MDB (M_{Out}). Liquid streams #451, #820, and #551 generated by the ANS, OTM, and ENS, respectively, would require measurement of agent concentration and total effluent volume in these streams. Measurements of agent concentration and volume would also be required for the gaseous effluent in stream #807 emanating from the OTM. However, unlike liquid effluent from the ANS, ENS, and OTM, the gas-phase streams cannot be impounded, which means that sampling would need to occur in-process. This also means that the gaseous streams cannot be reprocessed for additional treatment if necessary. These factors will further complicate measurement of M_{Out} under Approach 3 (see below).

Finally, residual agent mass on the metal generated by the MPT (stream #7652) would need to be measured, although it may be possible to replace measurement with process knowledge based on previous experience that has shown that metal parts subjected to high temperature treatment (1,000°F) for 15 minutes contain no agent. Residual agent mass on the particulate matter from the OTM (stream #804) would also need to be measured, or deemed zero based on process knowledge. Summing the agent masses from these solid streams, together with the liquid stream #820 and the gaseous stream #807 exiting the MDB, would provide a valid M_{Out} value.

A concern with Approach 2 is that it does not account for any agent that survives the EBHs and escapes the OTE to the MDB HVAC system. This could occur because the vapor pressure of GB is high at the operating temperature of the caustic in the EBHs, so volatilization may be competitive with hydrolysis. While the capacity of the carbon filtration system is likely adequate to capture fugitive agent that has escaped the OTE, the quantity of agent that might be partitioned in this effluent stream will not be known. If the state of Kentucky does not allow carbon capture in the MDB HVAC system to count for agent destruction, this factor could impact the ability of BGCAPP to meet the DE criteria.

Finding 3-8. Approach 2 is not an appropriate option for the calculation of DE. It is incomplete because it does not include the gaseous emissions from the OTE—which, under the new configuration, may contain agent—and because it is not operationally practical to measure agent quantities in the feed to, and effluent from, the individual process units.

Calculation of Destruction Efficiency Under Approach 3

Under Approach 3 the overall process flow for calculating DE is considered to have a single point of agent entry—that is, intact projectiles or rockets entering the MDB. Thus, it is not necessary to measure the agent masses, concentrations, or volumes at the entry points for each of the process units, because M_{In} values could be readily estimated, with a reasonable degree of accuracy, from process knowledge of the quantity of agent in each type of munition and the number of munitions to be processed per batch or per unit time. The partitioning of agent through the various units and their effluent streams is identical to that found in Approach 2, so all the considerations for measuring the contributions of the different streams to M_{Out} are the same, except Approach 3 accounts for the possibility that vaporized agent might be transferred to the OTE. The gaseous effluent stream from the OTE (stream #8517) could contain measurable agent and is sent to the carbon filter banks in the MDB HVAC system.

Summing the M_{Out} values produced by these streams, together with the derived M_{In} value, would be sufficient to calculate a rigorous and defendable DE value in accord with the KRS and RD&D permit requirements,[17] using the following modified DE calculation to account for the partitioning of agent resulting from the deletion of agent washout:

$$DE = 100 \times (M_{In} - M_{451} - M_{551} - M_{820} - M_{807} - M_{8517} - M_{804} - M_{7652} - M_x)/M_{In} \quad (4)$$

---

[16] RD&D permit.

[17] John McArthur, environmental manager, BPBG, "Destruction Efficiency Considerations," presentation to the committee on September 9, 2015.

where

M_{In} = the mass of agent input to the agent demilitarization area, presumably to be calculated as the product of the concentration and the volume or mass of the projectile or batch;

M_{451} = the mass of agent in the liquid effluent from the ANS, calculated as the product of the concentration and the volume;

M_{551} = the mass of agent in the liquid effluent from the ENS, calculated as the product of the concentration and the volume;

M_{820} = the mass of agent in the liquid effluent from the OTM, calculated as the product of the concentration and the volume;

M_{807} = the mass of agent in the gaseous effluent from the OTM, calculated as the product of the concentration and the volume;

M_{8517} = the mass of agent in the gaseous effluent from the OTE, calculated as the product of the concentration and the volume;

M_{804} = the mass of agent residual on the metal parts from the MPT;

M_{7652} = the mass of agent on the particulate from the OTM TOX unit; and

M_{x} = the mass of agent in the gaseous effluent from the PHS, MWS, RHS, and other fugitive agent vapor releases that are directed to the MDB HVAC.

This approach is shown in Figure 3-2. The practicality of this approach depends on the methods used to measure agent mass in these solid-, liquid-, and gas-phase streams. Measurement of agent concentrations and volumes in the liquid waste streams could be achieved using the approach

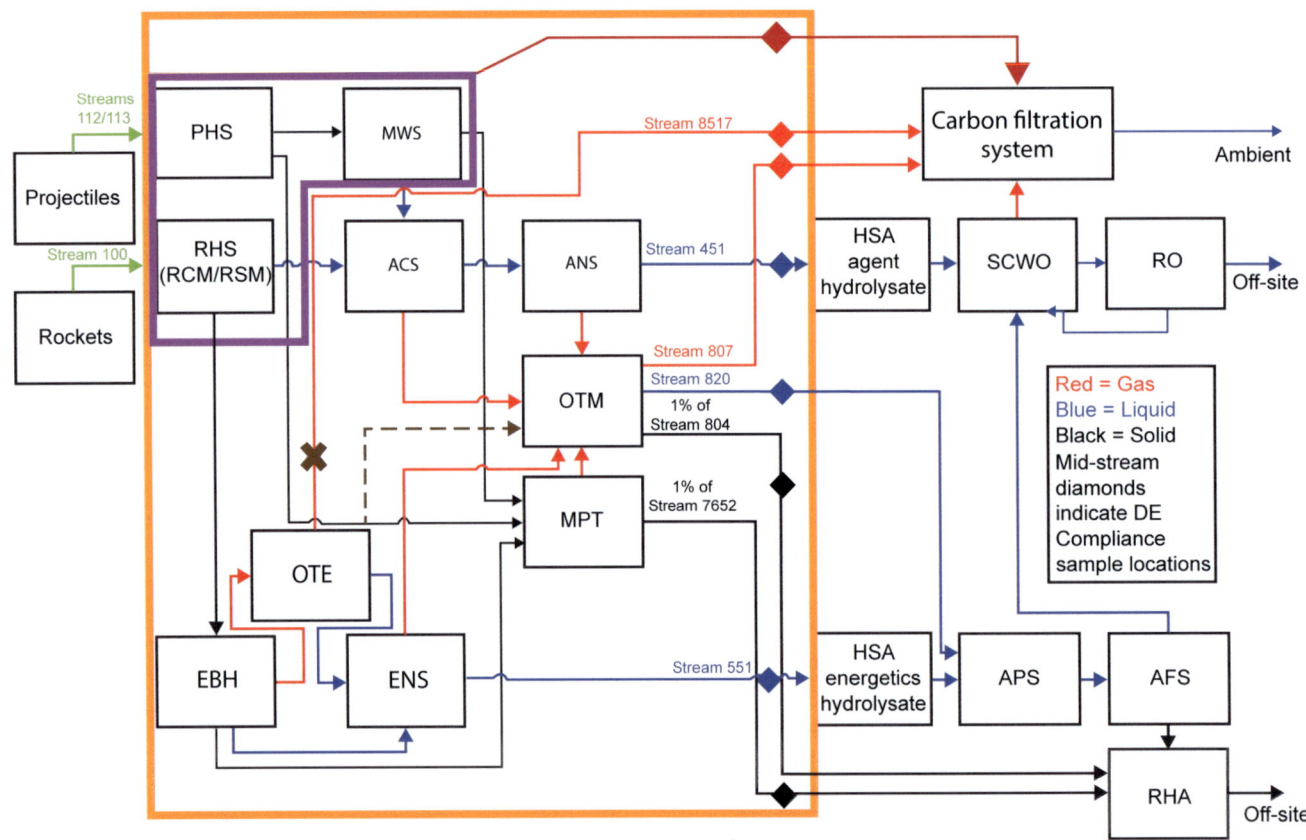

FIGURE 3-2 Flow diagram showing committee recommendations for expanding effluent measurements to allow the calculation of DE at 99.9999 regulatory requirements and for rerouting the OTE through the OTM. The orange and purple boxes represent the committee's interpretation of the measurement of BGCAPP effluents to be used in calculating the DE of 99.9999. The purple box around the MWS, PHS, and RHS, and the unnumbered maroon line from the purple box to the carbon filtration system represent fugitive agent emissions from munition drain operations, which are sent directly to the carbon filter banks. The brown dotted line represents the committee's recommendation that OTE emissions be routed to the OTM; and the brown X on the red line directly out of the top of the OTE box represents the committee's recommendation to delete this stream upon rerouting to the OTM. NOTE: Acronyms are defined in the front matter. SOURCE: Adapted from John McArthur, environmental manager, Bechtel Parsons Blue Grass (BPBG), "Destruction Efficiency Considerations," presentation to the committee on September 9, 2015.

currently used for the ANS effluent. The quantity of residual agent on the metal parts from the MPT is expected to be negligible based on the temperatures and treatment times used and historical knowledge of agent destruction under these conditions. However, the amount of agent on the particles from the OTM will need to be measured. Moreover the gaseous waste streams from the ANS and the OTM will need to be sampled and analyzed for agent at levels sufficiently low to ensure that six-nines criterion is met. A drawback to this approach is that, if the gas-phase streams are found to contain agent, they will not have been held in containment and, thus, cannot be re-processed if additional agent destruction is needed. Thus, for both Approaches 2 and 3, the risk of not meeting the DE criteria is increased due to the uncertain amounts of agent that will be partitioned to the EBH and ENS/ENR, the potential fraction of this agent that will vaporize, the potential fraction of vaporized agent that will be sent to the OTE, and the timescales over which a process or facility response to these events must occur. Such details have not yet been finalized by BGCAPP.

Finding 3-9. Approach 3 could conceivably be used for a defendable DE determination, because it accounts for the OTE gaseous process stream #8517, provided it is modified to include the fugitive releases of agent vapor directed to the MDB HVAC system. However, Approach 3 would require development of additional methodologies for measuring masses of agent partitioned into the two gaseous waste streams entering the MDB HVAC system.

Recommendation 3-6. If Approach 3 is adopted, then BGCAPP should evaluate the concentrations of agent liable to be present in all gaseous process streams and develop measurement approaches with sufficient sensitivity to ensure that destruction efficiency criteria are being achieved.

Approaches 1, 2, and 3 for calculating DE; the potential for increased agent in the process streams after washout deletion; and the committee's recommended approach to calculating DE with rerouting the OTE off-gas through the OTM are summarized in Table 3-1.

TABLE 3-1 Summary of Approaches to Calculating Destruction Efficiency (DE)[a]

Stream	Source	Phase	Approach 1 (Original Design, Including Washout)	Potential for Increased Agent (No Washout)	Approach 2 (No Washout)	Approach 3 (No Washout)	Committee-Recommended Approach (OTE Routed to OTM, No Washout)
100	Input, rockets	n/a	Measure	N		Measure	Measure
112/113	Input, projectiles	n/a	Measure	N		Measure	Measure
n/a	Input, ANS	Liquid		N	Measure		
451	Output, ANS	Liquid	Measure	N	Measure	Measure	Measure
n/a	Input, combined MPT/OTM	Gas		Y	Measure		
n/a	Input, combined MPT/OTM	Solid		Y	Measure		
807	Output, OTM	Gas		Y	Measure	Measure	Measure
820	Output, OTM	Liquid		Y	Measure	Measure	Measure
804	Output, OTM	Solid		N	Measure	Measure	Measure
7652	Output, MPT	Solid		N	Measure	Measure	Measure
n/a	Input, ENS	Liquid		Y	Measure		
551	Output, ENS	Liquid		Y	Measure	Measure	Measure
n/a	Input, OTE	Gas		Y			
8517	Output, OTE	Gas		Y		Measure	
n/a	Output, ventilation from PHS, MWS and RHS (to HVAC)	Gas		Y			Measure

[a] Stream identification numbers are found in Figures 3-1 and 3-2. Stream measurement is identified for each of the DE calculation approaches. The potential for increased agent in the streams is also indicated. The final column addresses changes in measurement if the gaseous streams from the OTE are sent to the OTM.
NOTE: ANS, agent neutralization system; ENS, energetics neutralization system; HVAC, heating, ventilation, and air conditioning; MPT, metal parts treater; MWS, munitions washout system; PHS, projectile handling system; OTE, EBH off-gas treatment system; OTM, off-gas treatment system; RHS, rocket handling system.

MEASUREMENTS REQUIRED FOR VERIFYING DESTRUCTION EFFICIENCY REQUIREMENTS

As noted, one of the major limitations in calculating a valid DE after washout deletion is the increased number of measurements that will be needed compared to BGCAPP's original plan. Under its RCRA permit, BGCAPP will have to establish these analytical methodologies for measuring agent in the individual waste streams in its waste analysis plan and use such results to conduct the DE calculation for each initial batch of agent to be treated in the BGCAPP, as required under the KRS and the KAR.

The analytical measurement approach that was based on the original operational design to ascertain that the concentration met the six-nines DE criterion in the ANS liquid effluent will likely be applicable to the liquid waste streams from the OTM and the ENS under the modified operational design that no longer includes munition washout. However, clearance levels for the three liquid streams will vary depending on the fraction of agent partitioned into the ANS, MPT, and EBHs. At the present time, the fraction of agent that will be partitioned into these processing units is not known. However, a measurement of the concentration, together with a reasonable estimate of the volume produced per munition or per batch, will suffice to provide defendable M_{Out} values in the three liquid effluent streams. BGCAPP will have to determine whether the sensitivity of the current methodology will be sufficient to confirm that DE criteria have been met.

As stated above, it is likely that new methods for measuring agent on metal parts from the MPT will not have to be developed. This is based on the expectation that it can be reasonably demonstrated from historical data and process knowledge that agent subjected to the temperatures and residence times in the MPT will be completely destroyed. Unless it can be demonstrated that current methods are capable of measuring agent in the particulate matter generated in the OTM TOX, new methods will need to be developed for measuring that process stream.

The gaseous process stream from the OTM is also not likely to contain significant agent, based on historical experience of the DE in units similar to the TOX. However, the same statement does not apply to the gaseous effluent from the OTE. The fraction of intact agent that will be partitioned into the off-gas stream from the OTE is not known; however, the fraction partitioned will affect the analytical requirements for this stream and the methodology eventually settled on.

Finding 3-10. The performance requirements for the analytical measurement methodology for measuring agent in the off-gas process stream from the OTE (#8517) are not known, because the fraction of agent that will be partitioned into this stream is uncertain.

Recommendation 3-7. If Recommendation 3-4 is not pursued, BGCAPP should conduct research to determine what fraction of GB agent might partition into the off-gas process stream from the OTE (energetics batch hydrolyzer off-gas treatment system) and then use this information to set analytical performance requirements that can be used to identify analytical measurement methodology.

Note, the committee considered whether a revision to the DE equation in the KAR would be possible that would allow for only measuring the DE at the final exhaust of the MDB HVAC system—in essence, including the removal of agent onto the carbon filters as the final step in treatment (i.e., destruction and removal efficiency). However, the committee could not determine the ability of BGCAPP to predict that the agent in the final exhaust from the MDB HVAC system would always meet the statutory requirement that the treatment or destruction technology has been demonstrated as effective in order to provide assurance of destruction or neutralization at an efficiency of 99.9999 percent for each compound under all operating conditions.

4

Process Modeling in Support of Washout Deletion

The Blue Grass Chemical Agent Destruction Pilot Plant (BGCAPP) is a one-of-a-kind facility that contains a number of first-of-a-kind (FOAK) pieces of equipment and procedures for the processing of nerve agents stored at the Blue Grass Army Depot. There is no standard of operation to compare against. BGCAPP is essentially a FOAK plant with a number of FOAK pieces of equipment. There is an inherent risk to any one-off chemical plant. In industry there is normally a pilot plant preceding the full-scale production plant. In this case, that is not really possible. At this writing, each part of the process has either been approved by the Kentucky Department for Environmental Protection in the current Research, Development, and Demonstration (RD&D) permit, or is being reviewed by the Kentucky Department for Environmental Protection (i.e., the design and process revisions presented in the current RD&D Permit Revision 5 Submission). Given the obvious risks involved with processing chemical warfare agents, it is important to use modeling to identify operational, management, maintenance, and safety issues rather than waiting to encounter them during actual operations. To this end, a discrete-event Monte Carlo simulation model was used by BGCAPP program staff to predict how long BGCAPP will be in use.

One should recognize that the BGCAPP model is a process model, not a model of the underlying chemical methodology of agent destruction. The process model assumptions describe the flow of material within the processing system. The system parameter values used in the BGCAPP model are based on estimates by BGCAPP personnel of chemical agent inputs and outcomes.

THE BGCAPP FACILITY MODEL

The BGCAPP process model was presented to the committee in block diagram flow format, shown in Figure 1-1. Comparing the structure of the diagram to the technical information presented to the committee at the initial meeting at BGCAPP, the model appears to represent the process well.

Given the complexity of the BGCAPP facility, it was difficult to determine from the committee's quick walk-through of the facility whether or not the structure of the model reliably represents the BGCAPP facility from a process-flow standpoint. It is clear, however, from presentations given to the committee and the ensuing discussion, that BGCAPP management supports the model from a structural standpoint and is fully invested in it. In fact, the model structure has been vetted periodically over time by BGCAPP personnel.[1] Given this, the committee's two main questions were the following:

- Was the model exercised appropriately? and
- Was the model fed appropriate parametric data?

Discussions with BGCAPP staff during the first meeting and later by telephone revealed that some of the input data for the model was gleaned from historical data that was gathered in a way such that the true stochastic nature (or random behavior) of the data was not identified.[2,3] Some of the input data were estimated by gathering all of the subject-matter experts in the organization together to make educated, consensus estimates of the values to be used in the model. Thus, it appears that most of the stochastic nature of many important system parameters is not available to the model. This should not be construed as critical of the processes that BGCAPP has used to estimate the operational data to be fed to the model. The data used in the model were simply the best data available at the time. The average processing rates and times used in the model could be accurate in the mean

---

[1] John Barton, chief scientist, Bechtel Parsons Blue Grass (BPBG), conference call with committee member Thom Hodgson on September 30, 2015.
[2] Ibid.
[3] John Barton, chief scientist, BPBG, Charles O'Classen, throughput engineer, BPBG/Bechtel Pueblo Team (BPT), Michael Noyes, and John Coyne, Bechtel Parsons Group, conference call with committee member Thom Hodgson on October 1, 2015.

(for example, the estimates of the time to drain the agent from weapons).

The proposed elimination of the washout functions will impact the operation of downstream treatment systems and thus the model input parameters. Two processes that will be impacted are presented as examples below: munition drain times and filter sock change-out activities.

Finding 4-1. While the process model explores the influence of variations in operating parameters on the performance of BGCAPP, the limited treatment of the stochastic nature of those parameters does not reflect operational experience.

Finding 4-2. The reliance on point estimates in the model data does raise concerns about the ability of the model to accurately forecast future facility operations in terms of the length of time to complete the processing of the chemical weapons and the risks involved in operating the facility.

Munition Drain Times

With washout deletion, there may be significant differences in munition drain times that the model is not able to reflect because of the non-stochastic nature of the data fed to it. In the case of the initial drainage amounts from the munitions, the committee spoke with individuals who had actually drained a number of similar munitions 10 or so years ago.[4] Anecdotal accounts indicated considerable variance in the condition of the GB agent in the weapons. They commented on observing the state of the agent in some cases as gelled and/or crystallized. It is certainly unlikely that the condition of the munitions has improved over the past 10 years. This informal data might be used to estimate (at least) upper and lower bounds on drain time. Running the model using the upper and lower bounds would give a range on the actual time to complete agent processing. Were data available on the distribution of drain times for particular munition types and agent lots, the model predictions could gain even greater fidelity with actual plant operations.

Finding 4-3. The stochastic nature of the gelling or crystallization of the GB agent may still be partially retrievable. A formal debriefing of individuals who have drained munitions to capture the (informal and clearly anecdotal) nature of the condition of the agent in the weapons might be useful in developing more believable assumptions as to the condition and variability of the chemical agents in the weapons.

Recommendation 4-1. BGCAPP should retrieve and document historical (informal and anecdotal) data on munition drain times and run these data, complete with ranges of uncertainty, through the BGCAPP model.

Filter Sock Change-Out Frequency

The BGCAPP model currently assumes filter sock change-out once every 3 days.[5] This was said to be a conservative estimate that is based on assumptions about how much gel will be in the munitions, how much of that gel will drain from the munition using gravity, and how much will be trapped in the filters. Other than scheduled and unscheduled maintenance of equipment, the filter change-out rate may be the most critical step in terms of impact to schedule. That is to say, it may be the most important "pinch point" in plant operations that can impact schedule.

The main take-away point is that the length of time required to actually complete munitions processing at BGCAPP may have been underestimated using the BGCAPP model. This may have a negative schedule impact on BGCAPP operations, with implications for budget, treaty compliance, and the timely reduction of storage risk by destroying the stockpile.

Finding 4-4. The actual filter sock change-out rate may be the most important rate-limiting factor in BGCAPP operations and may be underestimated.

EXPLORING SYSTEM SENSITIVITY TO THE INPUT PARAMETERS

Many operational issues cannot be fully known until the facility is actually in operation. Nothing was presented to the committee relative to the sensitivity of the performance of the system to the various input parameters values.[6] It is clear that the BGCAPP staff has attempted to be conservative in all of their parameter estimates, but it is also clear that many of the parameters potentially have larger variances than expected.

The unavoidable deficiencies in the estimation of the system parameters used in the BGCAPP model, in and of themselves, argue for trying to estimate the sensitivity of the system to variations in the operating parameters. Considerable effort has been expended over time to validate the structure of the model.[7] Given the effort put forth, it can be expected that the point estimates used for the parameters are at least in the ballpark. However, the stochastic nature of the BGCAPP processes is not well represented in the model due to the reliance on point estimates.

---

[4] John Barton, chief scientist, BPBG, conference call with committee member Thom Hodgson on September 30, 2015.

[5] John Barton, chief scientist, BPBG, "Rocket Handling System/Munitions Washout System (RHS/MWS) Process and Infrastructure Changes Due to Washout Deletion" presentation to the committee on September 9, 2015.

[6] John Barton, chief scientist, BPBG, conference call with committee member Thom Hodgson on September 30, 2015.

[7] John Barton, chief scientist, BPBG, Charles O'Classen, throughput engineer, BPBG/BPT, Michael Noyes, and John Coyne, conference call with committee member Thom Hodgson on October 1, 2015.

As noted above, the model apparently does represent the process flow of the facility.[8,9] Thus, in its present form, the model should be sufficient to determine estimates of the sensitivity of BGCAPP operations to variations in the operating parameters. Bottlenecks can be identified as a function of varying various parameters in model runs. In this way, the potential for excessive filter cleanouts or excessive munition drain times to impede system performance can be explored. Other issues of importance might also be identified, explored, quantified, and mitigated (e.g., the need to process fewer munition bodies on the trays going through the metal parts treater).

In order to perform a sensitivity exploration exercise, it would be necessary to design a series of runs of the model—that is, initially placing relevant parameters at an upper bound and then at a lower bound (i.e., the highest and lowest levels). The objective is to determine a model of system responses to changes in the parameters (i.e., a response surface). Regression could be a reasonable way to develop a response surface from the model output.

Finding 4-5. Analysis of the sensitivity of the BGCAPP operations to variations in model input parameters might expose potential operational issues, allowing them to be quantified and possibly mitigated prior to operations.

Recommendation 4-2. BGCAPP should design and execute a series of modeling experiments to determine the sensitivity of operations to variations in operating parameters, reflecting the stochastic nature of some processes. Examples of parameters include maintenance and repair times, added characterization steps, retreatment for batches not meeting destruction efficiency, and compounding problems such as long munitions drain times together with very frequent filter sock change-outs. The results of these experiments should be used to prepare for potential challenges and mitigate them ahead of time as much as possible.

OPERATIONAL DATA COLLECTION

It is clear that the accumulation of data to characterize the actual values of operating parameters is the best way to model plant performance. For many critical parameters, however, this may be only possible to do as the facility enters operations. In other words, there may be no way, at this point, to improve current estimates of many of the operating parameters prior to actually bringing the system into actual operation. However, BGCAPP management plans to bring the BGCAPP facility into operation slowly and to verify predicted chemical reactions, reaction rates, and thermodynamics. This will provide the opportunity to collect data on critical operational parameters and develop parameter baselines for later operations. This start-up effort is critical to the operation of the facility and to updating estimates of the time that will be required to complete munitions processing. BGCAPP has developed a comprehensive control center to gather, in real-time, all relevant operational parameters, which would make gathering the appropriate data straightforward.

Early operations will be the first time that these parameters will, in fact, have the potential to be accurately estimated from actual operational data. However, it is important to realize that these estimates still contain randomness. With improved estimates of the parameters and of their statistical distributions, continued real-time forecasts of system performance can be made with the model to tune the processes, to improve the model, and to aid in the management of the overall system. Note that doing this will allow modifications to the model, as many parameters that are now modeled as fixed parameters will actually be able to be modeled as stochastic in nature. For example, data on drain times of a particular munition may fall into multiple classes, with some munitions draining rapidly and completely and others, where gelation or crystallization has occurred, draining more slowly and less completely. Such data could be represented in the form of a probability density function that would then replace the point estimates used in the early modeling.

Finding 4-6. Point estimates of operational parameters are only a starting point. To fully understand the plant operation and, ultimately, to understand the plant timeline, one needs data on the distribution of parameter values that may be encountered during operation.

Recommendation 4-3. During start-up, and continuing through plant operations, BGCAPP should gather data for relevant model parameters with sufficient resolution to assess the probability density functions for these parameters.

The stability of system operation will be important to observe and control. The concepts of statistical quality control could be of use in the analysis of the data (Grant and Leavenworth, 1974). This is a tool used regularly in industry to help control processes and maintain process integrity. Essentially, it is a methodology that allows statistical analysis of operating parameters to detect if a parameter is within operational bounds and to determine if the processing system is operating as required. The committee notes that BGCAPP is going to do some of this during systemization and operations ramp-up.

Finding 4-7. Statistical quality control could be a useful management tool for understanding and identifying possible problems as they occur.

---

[8] John Barton, chief scientist, BPBG, conference call with committee member Thom Hodgson on September 30, 2015.

[9] John Barton, chief scientist, BPBG, Charles O'Classen, throughput engineer, BPBG/BPT, Michael Noyes, and John Coyne, conference call with committee member Thom Hodgson on October 1, 2015.

Recommendation 4-4. BGCAPP should give attention to developing analysis tools such as statistical quality control prior to actual facility start-up.

REFERENCE

Grant, E.L., and R.S. Leavenworth. 1974. *Statistical Quality Control*. New York, N.Y.: McGraw-Hill.

Appendixes

A

Committee Activities

FIRST COMMITTEE MEETING
SEPTEMBER 9-10, 2015
RICHMOND, KENTUCKY

Objectives: To conduct introductory discussions and briefings (i.e. administrative actions, including committee introductions and composition/balance/bias discussions for committee members), discuss the Statement of Task and conduct a background review with sponsor, receive briefings and engage in dialogue with Blue Grass Chemical Agent Destruction Pilot Plant staff, review the report realization process and project plan, review and flesh out the Initial Report Outline, make committee writing assignments, and set future meeting dates and next steps.

Presentations

BGCAPP Process Overview, George Lucier, BGCAPP Deputy Chief Scientist

History Leading Up to Washout Deletion—Original Design and Reasons for Engineering Change, Neil Frenzl, BGCAPP Resident Engineering Manager

Process and Infrastructure Changes Due to Washout Deletion, Neil Frenzl, BGCAPP Resident Engineering Manager

Impacts of Washout Deletion on Metal Parts Treater and Thermal Oxidizer, George Lucier, BGCAPP Deputy Chief Scientist

Destruction Efficiency Calculations, John McArthur, BGCAPP Environmental Manager

Process & Throughput Modeling, Charles Oclassen, BGCAPP/PCAPP Throughput Engineer

SECOND COMMITTEE MEETING
OCTOBER 20-21, 2015
WASHINGTON, D.C.

Objectives: To discuss the report draft, discuss any additional data gathering needed, conduct report drafting, achieve a First Full-Message Draft, make committee writing assignments, and discuss next steps.

THIRD COMMITTEE MEETING
DECEMBER 1-2, 2015
WASHINGTON, D.C.

Objectives: To discuss report draft, conduct report drafting, achieve a Preconcurrence Draft, identify final tweaks, and discuss next steps.

VIRTUAL COMMITTEE MEETING
DECEMBER 17, 2015

Objective: Conduct final report draft review and achieve concurrence.

B

Biographical Sketches of Committee Members

GARY S. GROENEWOLD, *Chair*, is a scientific fellow in the Energy and Environment Directorate at the Idaho National Laboratory (INL), where he has conducted research in surface chemistry, gas-phase chemistry, and analytical measurement since 1991. His research has focused on determining speciation and reactivity of radioactive and toxic metals (U, Np, Pu, Hg) and of toxic organic compounds (including VX, mustard, and sarin). Prior to 1991, Dr. Groenewold served in line management at the INL and as the technical leader for the organic analysis group. Before coming to the INL, he worked in anticancer drug discovery for Bristol-Myers and conducted research in surface chemistry during a postdoctoral fellowship at Oak Ridge National Laboratory. He received a Ph.D. in chemistry at the University of Nebraska in 1983, where he studied ion molecule condensation and elimination reactions under the direction of Michael Gross. He has authored more than 130 research articles in these areas and has served on several ad hoc committees for the National Research Council (NRC).[1] He currently chairs the standing Committee on Chemical Demilitarization for the National Academies of Sciences, Engineering, and Medicine.

HEREK L. CLACK is a research associate professor in the Department of Civil and Environmental Engineering at the University of Michigan. Previously, he was an associate professor in the Mechanical, Materials, and Aerospace Engineering Department at the Illinois Institute of Technology (IIT). He received his B.S. in aeronautical and astronautical engineering from the Massachusetts Institute of Technology (MIT; 1987) and his M.S. (1997) and Ph.D. (1998) degrees in mechanical engineering from the University of California, Berkeley. Prior to joining the IIT faculty, Dr. Clack was an NRC postdoctoral fellow in residence at the National Institute of Standards and Technology in Gaithersburg, Maryland

(1998-1999), and a member of the technical staff at the Rocketdyne Division of Boeing Corporation (1987-1992). He is engaged in research and publication in the general area of transport phenomena within dispersions. In particular, his research addresses combustion of droplets and sprays and mercury emissions from combustion.

RICHARD C. FLAGAN is the Irma and Ross McCollum/William H. Corcoran Professor of Chemical Engineering and chair of the faculty at the California Institute of Technology (Caltech). The author of more than 340 journal articles and one book, Dr. Flagan holds 23 patents for developments in aerosol instrumentation and materials processing. His work has been honored by the Fuchs Award, jointly administered by the American Association for Aerosol Research, Gesellschaft für Aerosolforschung, and the Japan Associate for Aerosol Science and Technology; the Award for Creative Advances in Science and Technology of the American Chemical Society; and the Thomas Baron Award of the American Institute of Chemical Engineers, among others. Dr. Flagan is a member of the National Academy of Engineering (NAE). He received his B.S. in mechanical engineering from the University of Michigan in 1969 and his Ph.D., also in mechanical engineering, from MIT in 1973. Dr. Flagan joined the Caltech faculty in 1975 and developed an internationally recognized research program in environmental and nonenvironmental aerosols.

REBECCA A. HAFFENDEN currently serves part-time as a program's attorney at Argonne National Laboratory. Her recent professional work has included work for the U.S. Department of Homeland Security to evaluate legislation and regulations associated with security vulnerabilities and providing legal expertise to programs involving federal facility site remediation and hazardous waste compliance and corrective actions. Ms. Haffenden also coauthored a working paper on the application of federal and state hazardous waste regulatory programs to waste chemical agents, in addition to

---

[1] Activities of the National Research Council are now referred to as activities of the National Academies of Sciences, Engineering, and Medicine.

being a co-author of the Environmental Impact Statement for the Assembled Chemical Weapons Alternates program. She received a B.A. in psychology from the University of Illinois and J.D. from Suffolk Law School, Boston, Massachusetts.

THOM J. HODGSON is a Distinguished University Professor in the Edward P. Fitts Industrial and Systems Engineering Department at North Carolina State University. Dr. Hodgson's research has focused on scheduling and logistics. The problem areas run the gamut from classic job shop scheduling, to specific industrial scheduling problems, to supply chain issues, to military logistics and operational problems. Dr. Hodgson is a member of the NAE, the Institute of Industrial Engineers, and the Institute for Operations Research and the Management Sciences. He earned his B.S.E. in science engineering in 1961, his M.B.A. in quantitative methods in 1965, and his Ph.D. in industrial engineering in 1970, all from the University of Michigan.

MURRAY G. LORD is director of Environmental Health and Safety (EH&S) in the EH&S Operations Technology Center at Dow Chemical Company. He is responsible for research program for technology development for Global Environmental Operations, which includes project areas in process optimization, technology development, and capital project execution. Mr. Lord has experience in project areas across multiple business and technology areas. He is also accountable for EH&S performance, budget performance, project development, and personnel leadership of research group from four locations, and he is the leader of the Environmental Technology Leadership Group, which is accountable for environmental technology development for Dow. Previously, Mr. Lord was a technical leader of Propylene Oxide Process Research and was responsible for research program in support of technology development of the propylene oxide process. He was also responsible for development and coordination of research studies at laboratory, pilot plant, and full commercial scale.

WILLIAM J. WARD is currently a retired research engineer. He joined the General Electric Corporate Research and Development Center in 1965, where for 10 years he worked full time in the area of membrane gas separations. In subsequent years he worked part time with GE and other colleagues on membranes. He did pioneering work on facilitated transport in immobilized liquid membranes and on ultrathin polymeric membranes. The latter resulted in a medical oxygen-enrichment appliance. Dr. Ward was a manager from 1976 to 1979, after which he resumed full-time research activities in the area of catalysis. His catalysis work in the 1980s provided new understanding of, and a much-improved catalyst for, the chemical reaction that is at the heart of the silicone polymer industry. From 1990 until 1995, Dr. Ward worked on understanding and improving the performance of polyurethane foam insulation and on solving problems associated with the elimination of chlorofluorocarbons as foam blowing agents. From 1996 through 1998, he was the technical leader of a team that made another major advance in the synthesis of silicone polymers. In his last 3 years at GE, he was involved in a successful effort to develop a manufacturing process to produce a ceramic metal halide lamp. After retiring from GE in 2000, he has consulted for GE and other companies, including those associated with Walter Robb. In 1987, he was elected to the NAE. Dr. Ward has 29 publications in refereed journals and 40 patents.